GEEKSPEAK

GEEKSPEAK

How Life + Mathematics = Happiness

Dr Graham Tattersall

FOURTH ESTATE · London

First published in Great Britain in 2007 by
Fourth Estate
An imprint of HarperCollins*Publishers*
77–85 Fulham Palace Road
London W6 8JB
www.harpercollins.co.uk

Visit our author blog: www.fifthestate.co.uk

2

Hand lettering and illustrations by Geoff Westby 2007

A catalogue record for this book is
available from the British Library

ISBN 978-0-00-726338-7

Typeset in PostScript Minion and Frutiger by
Rowland Phototypesetting Ltd, Bury St Edmunds, Suffolk

Printed in Great Britain by Clays Ltd, St Ives plc

All Fourth Estate books are printed on 100% recycled paper
with Blue Angel environmental accreditation.

Blue Angel is one of the world's first and most respected
environmental accreditation programmes. This high-quality
recycled paper, made from 100% post-consumer waste, has
been manufactured according to stringent controls.
For more information visit www.blauer-engel.de

*This book is written in memory of my father,
who helped raise six children as well as quietly going
through life inventing, analysing, and mending
anything and everything.*

CONTENTS

CONTENTS

PHOTOGRAPH ACKNOWLEDGEMENTS

All photos, unless stated, taken from Stock.XCHNG.com.

Introduction: Living with Numbers: #347053, Aleksander Milosevic, Belgrade, Serbia

1. Scrabbling for Words: #650192, Steve Woods/Pinpoint, Essex, UK
2. Pumping Iron: #313419, Josef Lluis Caldentey, Spain
3. Safe as Houses: Author
4. Well Connected: #368906, Cory McKenzie, Canada
5. Fatal Attraction: #696930, Marcos Santos, Brazil
6. Home Alone: #656412, Liv Crazy, Australia
7. Beam Me Up, Scotty: #757924, Artiom, Ukraine (StockExpert)
8. Hidden Death: #97469, Ken Kiser, US
9. It's For You: #6926, Stephen Gibson/ BudgetStockPhoto.com, Australia
10. War Chest: #522105, Marcin Barlowski, Poland
11. When the Wind Blows: #739813, George Bosela, US
12. You're Rubbish: #315128, Liton Ali, London, UK

PHOTOGRAPH ACKNOWLEDGEMENTS

13. Nappy Crap: #560650, Pascal Montsma, Netherlands
14. The Cunning Fox: #769992, Lynne Lancaster (weirdvis), UK
15. Fly Wheels: #735390, Asif Akbar, India
16. Bus Stop: #198141, Tomasz Kowalczyk, Poland
17. Stirring Airs: #442256, Steve Ford Elliot, Eire
18. Dream Flight: #748302, Martin Brooks, UK
19. Processing Power: #370098, Max Brown, Sydney, Australia
20. Soul Mates: #264203, Paul Preacher, London, UK
21. Idiot Calculus: #282161, Anka Draganski, London, UK
22. The Ghostly Present: #708887, Robert Rosmond, New Orleans, US
23. Bad Breath: #398156, Tudou Mao, Shijiazhuang, China
24. Wave Power: Author
25. Mechanical Neurosis: #242268, Philippe Ramakers, Belgium
26. The Final Judgement: #19734, alxm (StockExpert)
27. Heavenly Body: #707105, Wojtek Wozniak, Zabrze, Poland
28. Passing Water: #633517, Emin Ozkan, Zmin, Turkey
29. The Man in White: #700514, Joan Koele, Netherlands

GEEKSPEAK

INTRODUCTION
LIVING WITH
NUMBERS

How much can you work out about your own world?

For as long as I can remember, I've been fascinated by how things *work*. There's a whole world of cogs, circuit boards and equations behind familiar objects and events, or the figures and statistics we see in the media. I can pinpoint the exact moment when this fascination was born: it was as an eight-year-old child on a family trip to the Scottish Highlands. Frantically looking for a way to distract his squabbling children, my father announced that 'The first person to tell me the weight of that mountain gets to sit in the front of the car.' I took on the task, excitedly shouting out wrong answers.

This was power – working out a new number, a new piece of knowledge about the world that hadn't existed ten seconds earlier. And, it happened in a person's head. It was amazing.

And that was it: I became a Geek.

A family friend dropped by after our return. He was an amateur radio enthusiast, and his car was festooned with all sorts of antennas. 'Watch this,' he said through a large smile, and placed his finger at the bottom end of an antenna.

The faint, sweet smell of burning flesh drifted towards me as the current from the antenna flowed through his skin, heating and burning its extreme outer surface, but without the strength to cause any pain or real damage.

I was transfixed. How had that happened? Why hadn't it hurt? How deep had the current penetrated? And, was it possible to work out how much energy had been spent in his finger? Here were more questions, more knowledge to attain, more facts and figures, all calculable inside a person's head.

The fascination with the simple conjuror's trick of the burning skin has stayed with me, as a desire to explore the hows and whys lurking behind everyday life. And that, in a nutshell, is what this book is about: how to analyse your world and come up with your own new knowledge and understanding.

But perhaps most importantly, it's about your ability to judge and check 'expert' opinion for yourself rather than take it as granted – about using your numeracy to be better informed in debates about our shared future.

Have a look at a couple of 'expert' assertions. Do you believe them? How would you judge their validity?

> Britain should focus on the development of wind and wave power to meet its electricity generation needs.

> Hormone replacement therapy (HRT) greatly increases the risk of breast cancer.

Neither statement can be tested without using some numbers: the amount of power that can be generated by a wind turbine, and the electricity needed to run Britain; and the statistical significance and experimental controls used to make the HRT assertion.

But the numbers themselves are of little value. For a more complete knowledge of the issues, the assumptions, caveats and uncertainties in their calculation need to be understood. And the only way to understand all those is to do the calculation yourself.

Many people think that to perform that kind of calculation you need to be an expert in the given field – which is why we tend to rely on experts. You might go along with that, and decide that it's best to leave it to the professionals, the politicians and the people in white coats.

But there is another way: a path of knowledge, learning

and understanding – the way of the Geek. Not an obsessive, narrow-interest, malodorous Geek, but a nice, thoughtful, sweet-smelling Geek, the kind you'd like as a friend.

A true Geek can get at the truth by understanding the limits of a calculation, by knowing when to approximate 3.783 million as 'about 4 million'. A true Geek knows how to simplify the analysis so that the answer, though not spot on, will still be accurate enough to guide opinion.

Oh, and a true Geek is interested in the mathematics of body size, and of God, sex, food and a whole load of other things that also interest quite normal people.

And this is the rub. We live surrounded by figures: the power of a wind turbine, the amount the average family spends on food, the fuel consumption of a Boeing 747, the weight of sewage you create each year. Not being able to estimate such figures yourself – not speaking any Geek – is like living in a foreign country and not being able to communicate.

How can you trust statements made by academics, architects, scientists and politicians without checking the underlying numbers?

You can't!

But you *can* work it out for yourself. And what's more, once you've done that it's better than just knowing the fact. You'll understand the basis and limits of its truth.

So this book is about empowerment, by combining common sense, straightforward arithmetic and a little questioning of received wisdom.

This book shows you how to speak Geek.

1
SCRABBLING
FOR WORDS

How big is your vocabulary?

You know thousands of words with many different meanings. Jane Austen uses over 6,000 different words in *Pride and Prejudice*, and you can read them all without the slightest problem. In fact, your passive reading vocabulary probably exceeds 10,000 words. Can you remember the last time you heard or read a word whose meaning you didn't know?

On the other hand, your active vocabulary – the words you use in everyday speech – will be much more limited. On an average day you'll probably get by on a few hundred words. And those words say a lot about you: your sex, age and social class.

In the early 1990s, recordings were made of conversations and used to build a database of words in the English language. The database, held at the University of Lancaster, contains over 100 million words spoken by men and women of all ages and occupations. Some interesting facts were pulled out of this data by three researchers, Paul Rayson, Geoffrey Leech and Mary Hodges. One of the most startling is the difference between the kind of words used by men and those used by women. It turned out that there are certain words which act like fingerprints, showing that a conversation is between two men, or between two women.

These are the top three fingerprint words in women's conversation:

> she
> her
> said

And the three words most characteristic of man-to-man conversation:

fucking

er

the

Those top three female words are instantly recognisable as typical of 'girl talk'. Just eavesdrop on a conversation between two women chatting near the office coffee machine: 'And *she said* that *her* friend was really upset . . . And I *said* to *her* . . .'

As for the men, here are a couple of guys leaning over the open bonnet of a car: 'What's that *fucking* wire doing?' '*Er*, dunno. *The* battery's dead.'

Similar studies reveal words that distinguish social groups. Words such as 'actually' and 'really' are indicative of Groups A, B and C1; 'bloke', 'bloody' and 'pound' are distinctly C2, D and E.

The journey from the equality of baby burbling to speaking in ways that encode your gender, age and social status takes two or three decades, but you can go back to the first moments after your birth quite easily. Start by letting every muscle in your mouth and lips go slack. Now make a noise.

That grunt is called the *schwa*. It is the most basic, neutral vowel sound, and sounds similar, though not identical, when uttered by people with different mother tongues.

In the first few months after your birth, you'll start to babble, and by the time you're coming up to your first birthday you'll have a few words. Those words use vowel sounds such as *u* as in 'mum' and *a* as in 'man'. They'll be bracketed by primitive consonants or nasal sounds such as *m* and *n* to create important words such as 'mumma' and 'dadda'.

Fast-forward to the age of around eighteen months, and

you'll be making much more complex sounds by articulating most vowels and consonants, and introducing *l* and *n* sounds.

The extremes of the vowel sounds are the cardinal points of your language. In English, they range from *a* as in 'cat' and *i* as in 'hit', to *oo* as in 'hoot' and *aw* as in 'saw'. You can utter a kind of sound circle with the cardinal vowels. Voice them in sequence and you'll find that the sound changes smoothly from one vowel to the next.

Counting vowels, consonants, nasals, and *l* and *r* sounds, a fully developed English speaker can recognise at least forty-five basic sounds. They are called *phonemes*.

It used to be thought that each phoneme was a distinct acoustic event, but it is now accepted that many are psycho-perceptual. For example, the stop consonant *pp* in the word 'apple' does not exist by itself. A stop consonant is the sound we think we hear ourselves saying when we use our mouth to rapidly stop or start a sound. The *pp* in 'apple' is the sound made when we quickly close our lips to stop making the *a* sound, and then explosively open them again to continue with the *le* sound.

We perceive the stop consonant as an actual sound that exists between the *a* and the *le*. But if you look at the sound trace of someone saying 'apple', you'll see a period of silence in the middle of the word. That's the *pp* in 'apple'. It doesn't exist: it's simply perceived because of the way the *a* and *le* are stopped and started.

One of the drawbacks of growing up speaking your mother tongue is losing responsiveness to speech sounds in other languages. This was demonstrated by Japanese researchers. They played sounds to infants while monitoring the frequency with which the infants sucked on a dummy. They sucked more often when there was a recognisable stimulus such as their mother's face or a familiar sound.

Newborns, who had barely been exposed to their mother tongue, sucked rapidly when they heard any of a wide variety of sounds drawn from many languages. Older infants sucked rapidly only when they heard sounds used in their mother tongue. The researchers inferred that infants lose the ability to distinguish certain sounds when they start to learn a language in which those sounds are absent.

Now, hopefully many years after you stopped sucking because something seemed familiar, you understand, speak and read thousands of words. It's rather strange that we bother, when just a few hundred words are sufficient for our daily lives.

So, how many words do *you* know?

It's possible to work out the size of your passive vocabulary. One approach is to go through every entry in a dictionary and tick off every word you know. But if you've got other things to do in life, there's a smarter way that gives a good estimate in a much shorter time. That method is called statistical sampling.

The idea behind statistical sampling is the same as used in surveys of, for example, voters. The nation's intended voting pattern could be found by asking all 30 million voters about their plans for the voting booth. More practically, a representative sample of voters is questioned. The sample might consist of just 1,000 people carefully selected to represent all the localities and social groups in the country.

The same approach can be used to estimate your vocabulary. Sample the 'population' of words by opening the dictionary at random 100 times. Each time, look at the first entry at the top of the page. Do you know the meaning of this word? If the answer is yes, add one to your word score. At the end of the exercise, divide your score by the sample size of one hundred to get an estimate of the fraction of words in the dictionary that you know. Multiply that fraction by the total number of words in the dictionary to make an estimate of your vocabulary size.

This method works, but you need to be careful: how many times should you dip into the dictionary at random to get a good estimate? Say you do the test twice and find that you know the first word, but not the second. That means that you know 50% of the words in the tiny bit of the dictionary you examined.

But common sense tells you that this estimate is unreliable. It is true that you might know half the dictionary, but it is also possible that you know 10% or 90% of all the words. The two words you chanced upon might have been unusually uncommon, or unusually common. Two out of however many thousand words the dictionary defines is not a representative sample.

Do the trial 10 times, and confidence in the result is greater; 100 times, even better. If you did the trial 1,000 times and found that you knew 500 words, you could argue quite strongly that you really do know about half of all the words in the dictionary.

To complete the estimate of your vocabulary you'll need to know the total number of words in the dictionary – preferably without having to count them. This is quite easy: look up the number of the last page in the dictionary, and take that as the number of pages. Next, open the dictionary at random and count the number of different words listed on that page. Multiply the number of pages by the number of words per page, and you have an estimate of the number of words in the dictionary.

I thought I'd better test myself using this statistical sampling technique. The dictionary I used has about 60 entries on each page, and over 800 pages. That's around 48,000 words altogether.

I opened the dictionary 125 times, and made a tick on a piece of paper if I knew the meaning of the word at the top of the page, and a cross if I didn't. Like me, you'll probably find it hard to stop yourself jumping ahead to other entries if the first is unfamiliar. Don't – that's cheating, and invalidates the statistical sampling!

The result: there were 25 words whose meaning I didn't know. On that basis, my passive vocabulary is 48,000 multiplied by 100/125. That's around 40,000 words. It sounds high, but it includes all the possible extensions of the stem of each word. For example, take the word 'abstract'. The dictionary will include 'abstractedly', 'abstractedness', and so on. The number of stem words I know is a lot less than 40,000.

Still, I'm feeling pretty good about myself, so I'm going to exercise my gigantic male vocabulary by introducing the next chapter:

'The, er, next chapter is, er, fucking interesting . . .'

SPEAK GEEK

'IT IS A TRUTH UNIVERSALLY ACKNOWLEDGED THAT A SINGLE MAN IN POSSESSION OF A GOOD FORTUNE MUST BE IN WANT OF A WIFE.'

Some authors are instantly recognisable from their vocabulary. For example, everyone recognises the style of Jane Austen, and many would say that her writing's distinguishing feature is its abundance of long words. But is this true? A bit of statistical analysis can reveal the answer.

The four longest words used by Jane Austen in *Pride and Prejudice* have 16 or 17 characters. They are 'superciliousness', 'communicativeness', 'disinterestedness' and 'misrepresentation'. But just looking at the longest words is not enough: we need to examine the distribution of word lengths over her entire vocabulary, as shown in the graph below:

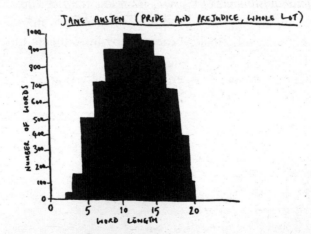

For comparison, here is the 'fingerprint' of the writer Ian McEwan, showing that his vocabulary includes many shorter words:

And, what about this book? *In this work I intend to speak with candour, and without misrepresentation or superciliousness, of the accomplishments of the irreproachable retrospections . . .*

2
PUMPING IRON

Are you as powerful as a washing machine?

For the price of a few cupfuls of oil, men can be transformed into mechanical supermen. Next time you pass major road-works or a large building site, watch the hydraulic rams on a mechanical digger pushing the bucket and scooping up one-ton heaps of sand, all at the twist of the wrist of the driver. Those rams become extensions of his limbs: he is a super-man.

Mechanical power is often quantified as *horsepower*, a word coined by the eighteenth-century engineer James Watt, the man whose work changed steam engines from profligate steam guzzlers into much more efficient and powerful machines.

In Watt's day, ponies or horses were used to turn a windlass that hoisted buckets of coal up a mineshaft. He would have wanted to know how many horses would be needed to lift a bucket in a given time. Watt knew that a horse could pull with a force of about 180 pounds, and that it could walk a total distance of around 180 feet each minute while pulling the load. That became his definition of horsepower: it's the power needed to move a force of 180 pounds through a distance of 180 feet every minute.

To get a feeling for one horsepower, think of it this way: an average man weighs around 180 pounds, so with a suitable pulley and rope, a horse could hoist him 180 feet into the air in about 1 minute. The Eiffel Tower in Paris is 986 feet high. If we could position our hoisting pulley at the top of the Tower, our man would be dangling almost one-fifth of the way up after 1 minute.

Connect the same pulley rope to the 60 hp engine in your car, and you could hoist the same man to the top of the Eiffel Tower in less than 6 seconds, although he might not have much stomach for the view once he arrived at the top.

What about men instead of horses? A simple way of measuring your power in horsepower would be to tie the pulley rope around your own waist, take the strain, and see how long it takes you to hoist the 180-pound man through 180 feet.

A 180-pound man is too much for most of us to lift, so let's replace him with a small child weighing, say, a quarter of that, 45 pounds. If you managed to hoist the child through 180 feet in 1 minute, you would have a power of one-quarter of a horsepower.

In reality, even if someone was willing to lend you their child in the service of science, most people would have difficulty in performing the task in less than 2½ minutes. So, a man's power is nearer one-quarter divided by two and a half, which is one-tenth of a horsepower.

Nowadays power is usually measured in watts rather than horsepower. We've just changed from using Watt's own term, horsepower, to using his second name as our standard unit of power. There are 746 watts in one horsepower.

Power comes in many forms, but it's always a measure of the rate at which energy is delivered somewhere. For a car it's the rate at which mechanical energy is delivered at the engine's flywheel. For a gas cooker it's the rate at which heat energy is delivered by the burner to the bottom of the pan, and for a light bulb it's the rate at which electrical energy is supplied to the bulb.

It would be quite legitimate, and possibly more meaningful, to rate, say, a 75-watt light bulb as one-tenth of a horsepower. A label of 1/10 hp on the bulb would indicate that one horse turning a windlass connected to an electrical generator could light ten such bulbs. More sobering, it shows that one athlete turning a windlass or a treadmill could manage to keep just one 75-watt bulb burning.

I'm guilty of sometimes having up to 300 watts-worth of light bulbs switched on in my house during the evening. In a pre-fossil-fuel era, I would have needed four slaves continuously walking a treadmill to keep them alight. But don't be too smug – to boil your 3 kW electric kettle you would need forty slaves.

The use of watts as the unit of power has probably contributed to the divorce between our understanding of machine power and of human power. Labelling the power of commonplace machines and devices in manpower instead of watts would keep us much more aware of the gearing provided to our lives by fossil fuel.

In a world without fossil fuel, the unit of power might be the 'slave'. Perhaps the obscenity of slavery disappeared only because we invented other means of getting cheap energy.

If a slave is equal to 1/10 hp, your car has a 600-slave engine, your water heater is rated at 40 slaves, and your fridge is about 1 slave. Yes, there is one slave pedalling a generator 24 hours a day, seven days a week to keep your food cool.

Machine	Power in watts	Power in slaves
Pocket torch	1	1/75
Phone charger	1.5	1/50
Portable FM radio	7.5	1/10
Digital radio	10	1/7.5
Low-energy light bulb	18	1/4
TV on standby	25	1/3
Modern fridge	75 (daily average)	1.7
Incandescent light bulb	75	1
TV switched on	75	1
Electric kettle	3,000	40
Small oil central heating system	15,000	200
Your car	50,000	600
Tractor digger	150,000	2,000
Pendolino train	4,500,000	60,000
747 jumbo jet	90,000,000	1,200,000

A few more examples are in the table, starting with the least powerful and working upwards. The cost of the energy delivered by any of these machines is phenomenally low. Imagine that someone offered to pay you to climb onto a treadmill to generate power to boil a kettle for a pot of tea. How much money would you expect to receive?

Most kettles are 3 kW, which is equivalent to 40 slaves. It will take about one and a half minutes to boil with that power input. With only you on the treadmill it's going to take 40 times as long. You will have to tread the mill for 1 hour to boil the water to make the tea.

If you get paid the current minimum legal wage, you'll get £5.35 when the kettle boils. Compare that with what you'd pay for electricity generated by fossil fuel or nuclear power – about 10 pence for each thousand watts for each hour. The kettle will use $3 \times 1.5/60 = 0.075$ kilowatt-hours, costing you about 0.75 pence. So human power, even at its cheapest, is about 700 times as expensive as using fossil fuel.

Even worse, after an hour on the treadmill you will need a change of clothes: you're going to need the washing machine. That means more work on the wheel. There is no escape.

Domestic washing machines have a motor with a power of up to 250 W, about one-third of a horsepower, which is more than twice the power that you could generate, even if you rigged the washing machine to some form of treadmill or pedalling system.

A typical wash cycle can take over an hour, and the motor's energy consumption will be about 0.3 kilowatt-hours, costing you just 3 pence. More significant is the electricity used to heat the water in the machine. Getting the water temperature up to 50 or 60 degrees Celsius will use over 1 kWh of electricity – three times the amount used for turning the drum.

Climb back on your treadmill again: you'll be there for twelve hours to heat the water to wash your clothes. And then you're going to need another change of clothes . . .

SPEAK GEEK

MORE THAN 60% OF THE ENERGY FROM BURNING
PETROL IN YOUR CAR IS WASTED.

There is an unassailable limit to the proportion of the heat
energy that can be converted into mechanical power by any
kind of engine. The unconverted energy is then dissipated: in
the case of a car, out through the radiator and in the hot gases
from the exhaust pipe.

To figure out that wasted energy, you take the temperature,
T_1, of the hot gases made by burning the fuel and subtract the
temperature, T_2, of the exhaust gases leaving the machine.
Then divide this difference by $273+T_1$ (don't ask – too geeky).
Multiply that by 100 to get the percentage of the energy that
could be converted to mechanical power. The wasted energy
is the difference between 100% and the number you just calculated. As a formula, it looks like this:

$$\text{Percentage of energy wasted} = 100 - \left(100 \times \frac{T_1 - T_2}{273 + T_1} \right)$$

For a car engine, T_1 is the temperature of gases in the
cylinder just after it has fired, and T_2 is the temperature of the
same gases after they have forced the piston down and are
pushed into the exhaust pipe. The formula will give you a
wastage figure of 60–70%; the same limit applies to the turbines in coal, gas and nuclear power stations. The clouds you
see billowing out of the big towers at power stations are not
smoke: they are formed from hot water vapour carrying away
the percentage of the fuel's energy that is wasted as heat.

3
SAFE AS HOUSES

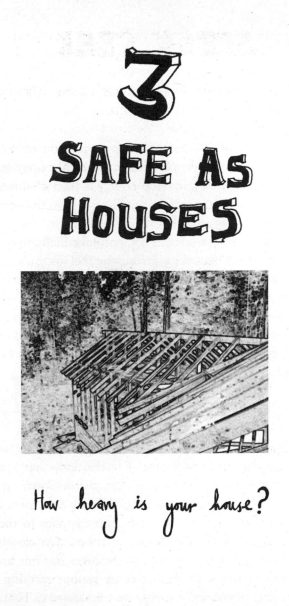

How heavy is your house?

Nowadays we all want to live in our own space. Children leave home in their late teens and want a place to live; newly-weds don't expect to camp with their in-laws; successful people want houses with offices, double garages and kitchens the size of a canteen.

Dealing with these changing demands requires new houses, and making the materials for building them takes a lot of energy. Oft-cited examples of energy profligacy – flying from London to Birmingham, leaving your TV on standby – pale by comparison.

There is a simple reason for this: building materials such as metal and bricks and concrete that are preformed into a useful shape are normally processed using heat. Heat means energy, energy usually means burning something which is often carbon based, and that means CO_2 emission.

Take concrete. There's probably tonnes of it under you as you read this book. The key ingredient of modern concrete is cement, a processed form of limestone and ash. Mix it up with some water, sand and stones, pour into a hole in the ground, and wait a few days. The slurry changes into a solid of enormous compressive strength that will support a tower block or a motorway bridge.

It sounds benign. But note that word 'processed' preceding 'limestone'. The limestone has to be heated in a furnace at the cement factory to change it into a substance that will combine with water to form solid concrete. The furnace is fired by a fossil fuel, usually gas or coke, and releases CO_2. Worse still, heating causes a chemical change in the limestone which releases even more CO_2 from the stone itself.

Manufacturing a tonne of cement pumps about three-quarters of a tonne of CO_2 into the atmosphere. That's the amount of CO_2 produced by burning 204 kg of carbon. If you had to go and buy the carbon as sacks of coal from your local shop, they would fill all the seats in your car, leaving only room for you to sit in the driver's seat.

The story is similar for bricks. Brick is made from clay which is fired in a furnace to make it strong and hard. Lots of energy goes into the furnace to get a usable building material, although with bricks there is fortunately no release of CO_2 from the clay itself.

Lightweight blocks are better. They're often made by taking the ash left over from burning coal at power stations and mixing it with a little cement and water. The mixture is then frothed up with air and poured into moulds to set. The blocks contain a lot of air bubbles, and just a little cement, which makes them much better than concrete in terms of expended energy. Lightweight blocks are true to their name – throw a dry block into water, and it will float.

But not all blocks are problem-free. Coal contains small quantities of uranium and thorium. Burning coal in a power station concentrates the radioactive material into the ash that goes into some building blocks. That makes them a little radioactive. Fortunately the level is very low, similar to the natural background from the slightly radioactive rocks in parts of Cornwall and Scotland.

Wood is used all over houses. It's light and cheap, and is ideal for structural components such as the rafters and joists that support roofs and floors. It can be shaped by a hi-tech, computer-controlled wood mill in a factory, or by a man with a half-inch chisel and wooden mallet. Most appealingly, it requires little energy to get it to its final shape.

The wood in a tree is made almost entirely of the chemical

elements hydrogen, oxygen and carbon, the last two of which come mostly from CO_2 absorbed from the air around the tree. A solid tree has been made literally just from thin air and water.

A living twenty-metre pine weighs up to a tonne, so each tree ties up a lot of CO_2. You might think that building with timber would deplete this natural CO_2 sink. But take a walk around the timber yard at your local builder's merchant. You'll see that the timber has its place of origin stencilled onto the side of the stack and the ends of each piece of timber. This is often the name of a Scandinavian or Baltic country, such as Sweden, Finland or Latvia. Those countries plant more trees each year than are felled, so using their timber encourages reforestation.

What other materials are used in building a house? There is lightweight stuff like copper cables and pipes for the electrical and water systems. Making copper uses energy, but the copper in your house doesn't weigh much so it doesn't contribute much to the building energy account. That isn't to say that it's cheap: in 2006 the cost of copper doubled because of the frenetic industrial development in China. Some small shops no longer sell electrical cable because it's too difficult to track the rapidly varying price.

The last significantly heavy and energy-intensive material in houses is steel. There may be steel reinforcing rods in the concrete foundations and in the cast concrete lintels over windows and doors. Sometimes steel beams are used to support floors and roofs.

Steel is no exception to the rule about heavy processed materials using a lot of energy. Changing iron ore into iron and steel requires heat. Most of that heat energy comes from coke and electricity. To produce 1 tonne of steel from ore takes about 3,000 kilowatt-hours of energy. That's the energy you would get by burning about half a tonne of coal; if you had solid-fuel central heating, it would keep your house warm for a month or two in winter.

The weight of a house is a good indicator of the energy used in its construction. So, how much does *your* house weigh? The answer will be about 100 tonnes, with two-thirds of that in the form of energy-intensive concrete and brick.

But how do you go about weighing your house? You could use a tape measure to get accurate dimensions and then work out the weight of each bit. But it's more comfortable to close your eyes, hold an image of your own house in your mind, and do some rough estimating from your armchair.

A useful mental tape measure is the length of a car. A car is 3 to 4 metres long, so judging how many cars could be parked along a wall provides an instant way of estimating the width and depth of a building. A similar mental trick to get building heights is to visualise how many men 2 metres tall would need to stand on top of each other to reach a given point on the building.

Start by estimating the outside dimensions. You might be able to park three cars bumper to bumper along the front of your house, and two cars along the side. That would make the front about 9 metres long, and the side 6 metres.

The height of the walls will be about the same as two and a half men standing on each other's shoulders. That would put the top man's waist level with the guttering, making the wall height 5 to 6 metres.

Now, with all the measurements in your head, the weighing can begin. Most houses are built on strip foundations of concrete. The strips are laid under every load-bearing wall in the building – the external walls, and also probably a dividing wall running up alongside the stairs.

The concrete is poured into foundation trenches about 1 metre deep and 0.6 metres wide, and its thickness will be at least 0.25 metres. Putting those numbers together, it's safe to say that at least 1 cubic metre of concrete is used for every 6-metre length of foundation strip. So, if the total length of load-bearing wall in the house were 42 metres, the volume of concrete in the strip would be 7 cubic metres.

That's not the end of the concrete. The entire ground area of the house within the footings will be covered with concrete to form a base for the floor. On top of that are waterproof membranes, thermal insulation and a cement screed, making the total concrete thickness in the floor about 150 mm. Its volume is the total thickness multiplied by its area, typically

8 cubic metres. That makes about 15 cubic metres of concrete in total, with a weight of about 30 tonnes.

Then there are walls. Modern houses have exterior cavity walls. That means an outer and an inner layer separated by a gap of around 60 mm. The gap helps to stop heat from escaping through the wall, and nowadays is usually filled with thermal insulation. The outer layer is often made from brick, and the inner layer from lightweight block. It's normal building practice to use blocks for the internal walls, so we'll count them separately.

The weight of the bricks is the number of bricks times the weight of one brick, which is typically 2 kg – the same as a couple of bags of sugar. The number of bricks in a house depends on the size of the brick, and to get the number we need to work out the area of the house walls and the number of bricks needed to make a square metre of wall.

Strangely, brick size varies from country to country, and in Britain it has even changed over time. In fact, British bricks have grown. For example, the brickwork in Hampton Court Palace looks a bit odd to our modern eyes: the bricks are only about 50 mm thick, whereas modern bricks are about 65 mm thick and 225 mm long. That works out to about 60 bricks in each square metre of modern wall.

The area of the average external house wall is about 150–200 square metres, so with 60 bricks per square metre you'll need around 10,000 bricks, with a total weight of 20 tonnes. That brings the weight of the house up to 50 tonnes.

The internal walls and the inner layer of the external cavity wall are made from lightweight blocks. The area of these walls depends on the layout of the house, but you won't go far wrong if you assume that the house is divided top to bottom by three internal walls running from back to front. The total area of those walls plus the inner layer of the external cavity wall will be about 250 square metres.

There will be about ten blocks per square metre of wall, so the house uses about 2,500 blocks. Each block weighs about 5 kg, giving a total weight of 13 tonnes (rounding up to the nearest whole number). The total weight of the house has now reached 63 tonnes.

Now, in your mind's eye, climb up on the roof and have a look at the tiles. They might be slate, clay or concrete. Many modern houses have concrete pantiles, but you'll have to check your own roof. A rule of thumb is that there are about 25 pantiles per square metre of roof, each weighing 2 kg; that's 50 kg per square metre. If the roof were flat, its area would be the same as the ground area of the house, but real tiled roofs have a pitch angle of at least 30 degrees to stop rain from blowing up under the tiles. That puts the roof area up by about 25%, to around 70 square metres, and the weight of all the tiles will be about 4 tonnes.

To complete the picture, you need the weight of timber and steel. Most big pieces of timber in a house are used for the floor and joists and roof rafters. They are usually spaced about 400 mm apart, so if you know the length of the house, the number is easily calculated. A figure of 30 is typical. The sizes of the joists and rafters range from 150×50 mm to 250×50 mm. Taking an average figure gives a volume of wood of about 6 cubic metres, which is about 5 tonnes in weight.

Lastly, there's the steel. The amount of steel in a house depends on its design. Some houses use almost none; others use steel in place of wood and concrete for structural components, or for reinforcing the foundations. It's quite easy to use 1 km of steel reinforcing rod in a complex foundation, so adding a couple of tonnes to the total weight of the house would be fair.

Altogether, the total weight of the house stands at 75 tonnes. But every builder knows that more material is needed than we have so far counted. It's something to do with the fiddly corners, stubs and overhangs that aren't shown on the drawings. Increasing the weight estimate by at least 25% will not be too much. That gives us a grand total for the house of around 100 tonnes.

So, if you're now alarmed about all those carbon-killing building materials, you may be wondering how many holiday flights could you offset by living in a tent. We'll make the numbers easy by assuming that your house has the equivalent of 50 tonnes of concrete in its structure. Making the concrete will release around 37 tonnes of CO_2 into the atmosphere. That's the same as 37,000 kg.

When you fly to Rome for a weekend break, the plane will burn aviation fuel at about the same rate as when you drive your car: 30–40 miles per gallon (10–14 km per litre). Of course, that's just for you and your baggage; the plane actually does about 0.2 mpg (less than 0.1 km/litre), but fortunately it holds more people than your car, so the consumption per person is much less.

Flying the round trip of 2,000 miles (3,000 km) to Rome and back will use 50 gallons (240 litres) of fuel on your personal account, and each litre of aviation spirit burnt releases about 2.5 kg of CO_2. Overall, your return trip will put approximately 600 kg into the atmosphere.

Will you forgo your new house for a tent so that you can have $37,000/600 = 61$ weekend breaks in Rome and still hold your carbon-neutral head up high?

SPEAK GEEK

AN AESTHETICALLY PLEASING WINDOW HAS A HEIGHT THAT IS 1.618 TIMES GREATER THAN ITS WIDTH.

Some houses look right and some look wrong. It often depends on the ratio of the length and width of the building's facade, and of the windows and doors and their positioning in the walls.

To many people, windows and other rectangular shapes look 'right' only when their width-to-height ratio is close to 1.618. This number is based on a mathematical rule called the Golden Section that has been used by everyone from the Ancient Greeks to Norman Foster. Using the Golden Section ensures an eye-pleasing harmony in the shape of the window.

Check your house-harmony by measuring the height and width of its window frames; divide the numbers, and give the result to your estate agent.

WELL
CONNECTED

Do you know the Queen?

When I was about seven or eight, I went to play with my elder brother on a railway track. My brother, deliberately winding me up, announced that he was going to derail the next train by placing an old penny on the rail. I believed him, and silently prayed for the imminent disaster to somehow be averted. Penny in place, we hid at the bottom of the embankment. The sound of the steam locomotive grew louder as it approached. There was a terrifying sound of steel on steel, of the snorting smoke box, and the train arrived . . . and passed by, oblivious to our presence. There were no newspaper headlines. The only evidence of our misdemeanour was the penny, now a flattened piece of copper about two inches by one inch, with a slight curl, lying quietly on the ballast between the rails. To this day I still feel a tingle of relief.

I tell this story because we, like other children of our generation, were pretty free of adult supervision. There were paedophiles, transformer yards and rail tracks waiting to catch the unwary child, but the sense of anxiety about strangers and the world's danger barely registered.

One cause of our present-day unease is maybe down to not knowing who is who in our community. In a village of three hundred people a hundred years ago, you would probably know which men had a predatory eye for children. What's more, you would know their name, and their parents and their brothers and sisters. Nowadays, you might not even know who lives two doors away in your street.

If we look at the number of children born to each couple and do some simple arithmetic, we can see why this has happened, and consequently why we are now faced with the prospect of identity cards, biometric devices and databases with information about our DNA.

The story goes like this.

How often has someone you know done some name-dropping by telling you about a friend of a friend who has met someone famous, or an eighth cousin twice removed who's starred alongside Tom Cruise? Maybe you know someone whose friend has a relative who works at Buckingham Palace and has spoken to the Queen.

I come from a middle-class family, but went to a secondary modern school. I have no obvious links to the aristocracy. But now I come to think of it, I do know a woman who was married to a man whose father was General Montgomery's adjutant during the Second World War, and who as a result was invited to tea with the Queen. That's just four links between me and the throne.

Try to forge those links for yourself, and you'll almost certainly succeed in finding a connection. In fact there's a good chance of making a chain of personal connections between you and anyone else on Earth using no more than six links in the chain. That means that a friend of a friend of a friend of a friend of a friend knows anyone you care to name – George Bush, Naomi Campbell, Osama bin Laden . . .

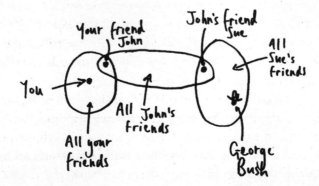

Osama bin Laden probably doesn't have mates that he meets in the pub, but he certainly has lots of other personal relationships that eventually lead back to you.

The fact that anyone on the planet is at most only six links away in a chain of relationships seems unlikely until you think about how many people you know on first-name terms. It's simple to estimate the total by dividing them into a few broad groups that can be enumerated. For example, family, friends, casual acquaintances and work colleagues probably cover most of the people you know by their first name.

Start with an average family in which parents have about two children. We'll assume that grandparents, uncles, aunts and cousins are included in the extended family. After all, you'll almost certainly be on first-name terms with all those people. The list will look something like this:

Relative	Number of people
Your mother's parents	2
Your father's parents	2
Mother and her sibling	2
Father and his sibling	2
Uncles'/aunts' partners	2
Cousins	4
You and your sibling	2
Total	**16**

So, if you are a child or young adult, there are probably around 16 members in your extended family. Now imagine going back two or three generations in Britain. A century ago, most couples had four or five children who survived into adulthood. It makes the table look dramatically different:

Relative	Number of people
Your mother's parents	2
Your father's parents	2
Mother and her siblings	5
Father and his siblings	5
Uncles'/aunts' partners	8
Cousins	48
Your siblings	4
Total	**74**

Evidently, in societies with large families it's the cousins who multiply almost exponentially with family size, and it is they who most likely provide the connectivity that holds communities together.

The reduction in the number of children in each family has had a profound and mathematically inevitable effect on community. In a small town of 10,000 inhabitants at the end of the nineteenth century, only a few hundred extended families would account for the entire population. You, or someone in your extended family, would be related to every other person in town.

Nowadays, with just 16 members in an extended family, you would have to live in a village of fewer than 200 geographically and socially immobile inhabitants to get the same level of coverage.

The tangible result of smaller families is low social connectivity and loss of community. Identity cards with biometric information and links to databases with our health records, past addresses and benefit history may be the inevitable synthetic replacement for first-hand knowledge about our neighbours.

A few years back, the Department of Trade and Industry organised a series of conferences on fraud prevention and security. They centred on the use of biometrics – technologies that use computer recognition of the iris, fingerprints or speech to uniquely identify an individual. One of those techniques is likely to be incorporated into British identity cards if they are ever introduced.

Grey suit after grey suit delivered the same technological gospel. Then, one speaker stood up and delivered a totally unexpected scathing attack on the whole idea. His argument: we have worked hard for our anonymity. It should be cherished as a prize of our development into a society based on individuals living without family. Who wants to go back to a village society where your every move is known to all?

Anyway, let's get back to analysing your connection to Osama bin Laden or the Queen. We've established that a typical extended family has 16 members, or 15 other people you know very well. How about work associates?

Assuming that they are people you know by first name, the number will depend strongly on the kind of organisation you work in. Self-employed people might know just half a dozen customers by first name, whereas someone working in a large insurance office might know 40 or 50. We could take a typical number as 25.

And then there are your friends: most friends are known to

you by their first name. You put in a figure for yourself. I think I have about 20 friends who are not anything to do with my work.

So far, the total number of people with whom you have a sufficiently close relationship to call them by their first name stands at about 60. That figure doesn't include all the casual acquaintances whose name you might know: the man at the post office, the window cleaner, the man you meet walking his dog. Let's say that makes another 20.

The total number of people with whom you are on first-name terms now stands at 80. Each of those people in your social circle is at the centre of their own personal social circle of about 80 people, and each of those people is at the centre of their circle. To simplify things, I'll use 'pal' to denote anyone who is a friend, a relative, a work colleague or an acquaintance.

As a start, we can say that the number of pals of pals we have is about $80 \times 80 = 6,400$. But in real life some of the pals' pals will be members of your own family, work colleagues and circle of friends, and the figure of 6,400 has counted those people twice over.

A simple way to take that overlap into account is to introduce a factor that expresses the percentage of each successive social circle that is common to the previous group. A factor of 50% seems fair. It means that, of the 80 people in the social circle of one of your pals, 40 will also be in your own circle.

So now, very conservatively, each link in the chain of social circles multiplies the number of people connected to you by a factor of just 40, but successive multiplication builds up the numbers very quickly. Try it out for six links in the chain:

Links	Number of people
1	40
2	$40 \times 40 = 1,600$
3	$40 \times 40 \times 40 = 64,000$
4	$40 \times 40 \times 40 \times 40 = 2,560,000$
5	$40 \times 40 \times 40 \times 40 \times 40 = 102,400,000$
6	$40 \times 40 \times 40 \times 40 \times 40 \times 40 = 4,096,000,000$

The table shows there are a total of 4.1 billion people accessible through a chain of just six links. The Earth's population is about 6.6 billion, so it seems that a pal of a pal of a pal of a pal of a pal will connect you to three-quarters of humanity.

Does it really work for anyone? Finding a link to Osama might be tricky, although certainly of interest to MI6. Bin Laden is a Saudi citizen, so a good bet would be to look for a link between yourself and someone in Saudi Arabia. Once you have found that link into the country, there are likely to be lots of links between your Anglo-Saudi bridge person and someone in bin Laden's family.

The Osama example shows how the small number of links could be exploited by anyone using information about phone calls or emails between individuals. The first step is to construct a database of all known names. Listed under each name are the numbers for all incoming and outgoing phone calls or emails for that individual. That's difficult, but technologically possible. Those phone numbers and email addresses are like the social circle of the named individual. A computer searches for chains with the same links.

We've already shown by simple arithmetic that there are probably no more than six links in the chain and that the people in the chain form a community of some kind. It may be benign or malign, but significantly it can be detected automatically by a computer.

It only takes one suspicious individual in the chain to cast suspicion on all the other people in the chain – the Queen, perhaps?

SPEAK GEEK

COUNTING FRIENDS OF FRIENDS, THERE ARE OVER A
THOUSAND PEOPLE IN YOUR SOCIAL CIRCLE.

A typical person has 30 to 50 people they consider to be close
friends or acquaintances. We could take an average of 40 for
the purpose of calculation. If each of these friends has a further
40 friends, excluding you and your shared acquaintances, there
will be approximately 40×40 people in your social circle.
That's 1,600 people. If you count friends of friends of friends,
the number rises to 64,000.

You'd spend a lot of time writing Christmas cards.

5

FATAL ATTRACTION

How much are you physically attracted
to your partner?

Have you ever woken up in bed with a numb arm or leg? The blood flow to the affected limb is constricted and leaves it feeling deadened. So you change position, the blood circulates again, and you go back to sleep.

The weight of a dead leg or arm seems enormous as you drag it into a more comfortable position. But how much do your individual body parts actually weigh? Your whole body weight is easy to measure: you just step onto some scales. Weighing a leg, an arm or even your head is more challenging, assuming that your body is to be left intact in the process.

Here is a practical method of working out an approximate value for the weight of your head while it is still connected to your body. You'll need a bucket large enough to immerse your whole head in, and a kitchen measuring jug. A bathroom towel might make the experiment more pleasant. It is best to do it in the garden, though be careful not to alarm the neighbours.

Fill the bucket to the brim with water and place it on the ground. Kneel next to the bucket, take a deep breath and gently immerse your head in the water, up to your neck. Water will spill out of the bucket, but this is quite normal.

When you've surfaced, the bucket will no longer be full. In fact, the amount of water that has overflowed will be exactly equal to the volume of your head. You can get a good enough measure of that volume by topping up the bucket with water using the measuring jug, until it is again full to the brim.

I just did this; you might notice the wet pages. The volume of my head I worked out to be precisely 5 litres. Now, I wouldn't like you to think that I'm being big-headed about this, but that makes my head the same size as one of those plastic petrol containers you keep in your car boot in case of emergency.

Hopefully you've also done the experiment and will now know the volume of your own head in the units of your measuring jug – litres, pints or fluid ounces. If the stuff in your head had the same density as water, the weight of your head would be equal to the weight of the same volume of water. A litre of water weighs 1 kg, so the weight of your head in kilograms would be the same as the number of litres of water. In my case (and for most adults) that'd be 5 kg.

Despite appearances, your body is mostly water. But compared with other parts of the body, a head contains quite a lot of bone relative to soft tissue, and so is going to weigh more than water, maybe 10–20% more. So just increase your initial head weight estimate by this amount. That brings it up to 6 kg.

In scientific terms, weight – whether of your head, your whole body or anything else – is a force that pulls between the mass of an object and the mass of the Earth. If you step onto old-fashioned bathroom scales that work by compressing a spring, the force that is your weight compresses the spring as surely as if you had squeezed it with your hands. What gives rise to that force is, of course, gravity.

Sir Isaac Newton worked out that the strength of the

gravitational force pulling two objects together increases in proportion to both masses, but decreases rapidly as the distance between the masses increases. He showed that the force was quartered for each doubling in distance.

We can write this as a formula:

$$\text{Force} = 6.67e - 11 \times \frac{\text{Mass}_1 \times \text{Mass}_2}{\text{Distance}^2}$$

That number, $6.67e - 11$, is called the *universal gravitational constant*. (The number written out in full is 0.0000000000667, but it is very tedious to write down all those zeros, and it's easier to use the shorthand built into calculators and used in computer programming languages. The -11 after the e indicates that the digits '667' are preceded by a decimal point and 11 zeroes. We can use the same notation to save writing out very large numbers in full: 4.567e9, for example, is 4,567,000,000, the age of the Earth in years.)

The first person to devise a method that could be used to get a very accurate value for the universal gravitational constant was Henry Cavendish. In 1798 Cavendish, reputedly a very shy man who may have had Asperger's syndrome, set up an experiment using an instrument called a torsion balance.

The torsion balance is a wooden rod suspended at its midpoint by a silk thread. Weights are hung from each end of the rod, a bit like the kind of mobile that people hang above a baby's cot. The delicate mechanism was sealed in a room so that it would not be disturbed by draughts.

Cavendish moved large balls of lead near to the outside walls of the sealed room, and they caused the suspended rod to twist very slightly because of the force exerted by the lead balls on the weights hanging on the torsion balance. The amount of twist can be used to work out the force and hence the universal gravitational constant.

That same gravitational force pulls on your body whenever there is a nearby massive object, for example a ship, a mountain or another human being. Unlike the pull from the Earth, those forces act along a line between you and the other object.

For example, when you stand right next to your partner, how much are you pulled together by the gravitational force between your bodies? Let's make it intimate – say you and your partner are standing face to face, just lightly touching.

It's simple to work out the strength of your attraction to your partner. You'll need some numbers to plug into the gravitational force formula.

Say that Object$_1$ is you. What is your mass? In numbers, it's the same as your weight in kilograms. This is just an estimate, so use a nice round number, say 100 kg. Next you need the mass of Object$_2$. That's your partner, and it's up to you to negotiate a value to insert here, but the calculation will be easier if you use 100 kg again. It may, of course, make other parts of your life more difficult.

Lastly, to use the formula you need a value for the distance between your bodies – and this is the interesting bit, because no single distance is correct. The distance at the points where your bodies touch, say, umm . . . chest to chest, is almost zero, but the distance between your buttocks and your partner's buttocks might be anything up to 60 cm.

The proper method of dealing with a problem like this is to imagine both bodies cut into very thin vertical sections, as if you had been put through one of those little utensils for cutting a boiled egg into slices.

Each slice would look like a paper cut-out silhouette of your body. You would use the area and thickness of each slice to get its mass, and then work out the distances between every slice of your body and every slice of your partner's body.

The gravitational forces between every pair of slices can then be worked out from the formula, and added up to give the total force of attraction. The mathematical process is called integration, and is easy to do on a computer, but too complex to do in your head.

What we need is a rough and ready method which will tell you quickly how much you are attracted to each other. He/she may be waiting impatiently to know the answer.

A simple solution is to assume that for you and your partner, all of your mass is concentrated at a single point in the middle of your body, say halfway between your belly button and the small of your back. That might be around 15 cm in from your front, making the effective distance between your body masses 30 cm.

And now you can run the formula. These are the values we need to plug in: $\text{Mass}_1 = 100 \, \text{kg}$, $\text{Mass}_2 = 100 \, \text{kg}$ and Distance = 0.3 metres. So your mutual attraction will be:

$$\text{Force} = 6.67e - 11 \times \frac{100 \times 100}{0.3 \times 0.3}$$

That's a force of about 0.0000074, and it is measured in units called newtons – after the great man himself.

Newton was involved in just about every field of mathematics and physics on which our modern lives are based. He formalised our understanding of gravity and force, coining the term 'gravity' from the Latin word *gravitas*, which means 'weight'. He invented differential calculus, and designed a type of reflecting telescope that is still called a Newtonian telescope.

Strangely, he got his knighthood not for his scientific advances but for his work as Warden of the Royal Mint. He stabilised the currency by linking the value of silver coins to the value of gold – a kind of precursor to the Gold Standard, used off and on until the Second World War.

He probably wasn't much interested in the kind of inter-body attraction we are calculating. He seems to have had only one girlfriend, whose affections dimmed as he immersed himself in science.

One newton is about the force needed to lift a 100-gram weight, so the force between you and your partner is only about the force needed to lift 0.000074 grams. Maybe Newton had already calculated that his force was too small to hold people together.

It is a measurable force, but you would be well advised to rely on the force of your personality, or other attractions, to keep your partner close.

SPEAK GEEK

A MAN WHO WEIGHS 100 KG AT THE NORTH POLE
WOULD WEIGH ONLY 99.65 KG AT THE EQUATOR.

Your net weight is the result of two forces. The first force is gravitational: it pulls your body mass downwards, towards the centre of the Earth. The second force is upwards, and is caused by the rotation of the Earth constantly trying to fling your body into space like a stone on a string being swung round your head.

At the exact point of the North Pole, the Earth experiences no rotation, and so there isn't any upward force. On the other hand, the Earth's rotational speed at the equator is at its maximum – about 1,500 kph – and a small amount of the Earth's gravitational pull is cancelled out by the upward 'flinging' force. Hence the weight loss.

Beat that for a diet, Professor Atkins.

6

HOME ALONE

How many piano tuners are there in Boston?

How much do you have in common with a Bushman in the Kalahari Desert? You both need food and shelter, shade from the sun when it's hot, and a source of heat for warmth when it isn't. Like him, you nurture your children. On the other hand, how often does a Bushman peruse the menu in a cappuccino bar?

We aren't all the same, of course, but each of the different societies that exist across the world can be placed in one of a fairly small number of distinct categories. Western industrial, agricultural, developing countries – these are labels that provide us with a convenient picture of a particular social, economic and cultural behavioural pattern.

Within those broad categories there are finer divisions that are culturally specific, and instantly understood by people belonging to a particular cultural group. If you're British, try dividing the list of words below into two groups with a common thread in each group:

Tuscany	Govan
satellite TV dish	pinstriped
Benidorm	Surrey
opera	cricket
comprehensive school	wages
pearls	hire purchase
Financial Times	director
football	salary
accordion	housing estate

Here is one obvious division based on social group:

comprehensive school	salary
football	pearls
satellite TV dish	opera
accordion	Tuscany
Govan	*Financial Times*
wages	cricket
housing estate	director
hire purchase	pinstriped
Benidorm	Surrey

We have strong preconceptions about the attributes of different social or lifestyle groups. They may be rooted in actuality, or be unfounded prejudice, but we use them all the time in an attempt to simplify the complexity of life.

In consumer societies, social labelling is often in terms of ownership rather than beliefs, income or occupation. Here's a sample of objects whose ownership is routinely used to put people into a nice tidy pigeonhole:

SUV	council flat
yacht	piano
hybrid car	garden
bicycle	hunting gun
second home	dogs
caravan	horses
detached house	swimming pool

Think about the things you own. How would you be labelled? For example, I play a piano accordion, and a lot of my friends don't regard the instrument as musically respectable. But probe a bit deeper and it becomes obvious that their judgement is more about association: the accordion is not a 'middle-class' instrument. And, looking around at other accordion players, it's true that many come from working-class backgrounds.

Accordions are portable, don't take up space like a piano and are relatively cheap – all things that matter if your income is low, your housing uncertain and you have to follow your job. And the accordion is the immigrants' instrument, allowing them to take a bit of the old culture to their new country. Immigrants start at the bottom of the social pile: a negative association for the accordion.

With pianos it's just the opposite. The piano is, after the violin and the cello, the most socially respectable of all instruments. Having a piano in your front room is a mark of intellect, aspiration and culture.

Social grouping and owning certain things go hand in glove. If you know the social group, you'll be able to predict the possessions; if you know the possessions, you can predict the social group. This relationship can be used to discover quite a lot about people and objects in places that you might never have even visited. All you need to know is something about either the social group or the possessions.

Take Boston, Massachusetts. Boston is a city in the developed world, just as British cities are. It features in TV shows such as *Cheers*, and the characters look very familiar to our British eyes. The city was founded by English puritans trying to escape an authoritarian state and Church in England, so its cultural roots are most definitely the same.

All in all, the social groups, value system and wealth of Boston are going to be similar to those of any moderately affluent English city, and the similarities allow you to make statistical predictions about the objects owned by people in Boston.

For example, if you knew the number of pianos, or accordions, owned by people in a British city such as Cambridge or Liverpool, you could directly estimate the number in Boston, by comparing the population of Boston with the population of Cambridge or Liverpool.

That's quite easy. Now try using the comparison method to predict a slightly less obvious quantity. Can you work out the number of *piano tuners* in Boston? Can you do it without looking in a book, phoning a friend, using Google . . . ? In fact, can you work it out sitting by yourself at home in your chair and with your eyes closed?

Even if you have been to Boston, the chances are that you weren't there to hire a piano tuner. But you can still make an estimate by assuming that the people who live in Boston lead a similar lifestyle to people around you. All you need to do is analyse the statistics of piano tuning where you live, and then apply what you've found to a city the size of Boston. Of course, the estimate will be poor if the locality in which you live is very different to Boston; for example, the ownership of pianos by flat-dwellers in Docklands will not be typical of Britain as a whole, let alone Boston. You need to allow for such variations.

Start with your own piano encounters. When you went to school you were probably in a class of about 30 children. Ask yourself this: how many of those children had a piano at home?

It's likely that there was at least one kid with a piano. On the other hand, it's not very likely that more than half a dozen children spent their evenings tinkling the ivories. Taking a middle figure, you might estimate that 3 out of 30 families have pianos. But this applies only to families with children at school.

When children grow up, some parents will get rid of the piano. So, if you want to extend your estimate of piano ownership to all households, you should probably reduce the original figure. Let's say 1½ pianos per 30 households instead of 3 pianos per 30 families – or in whole numbers, 1 piano per 20 households. (I'm saying that this figure applies across the country. If you think it's wrong, change it. It's your estimate.)

Now it looks easy. Just assume that the pattern of piano ownership is similar in Boston, estimate the number of households in Boston, and divide that by 20. Voilà – the number of pianos in Boston.

But remember, it's not simply the number of pianos in Boston we want, but the number of piano tuners. Also, another small point: how do you estimate the number of households in Boston when you are sitting in a chair with your eyes closed?

Let's deal with the second point first: how many households are there in Boston? We could get at this from a figure for the population by assuming that each household has an average of, say, three people. The number of households is then the population divided by three. Divide this number again by 20 to get the number of pianos.

All very simple so far, but we still need a population figure. You could just look it up, but the rules of this exercise forbid getting out of your chair. So the trick is to think of a city whose population you know, and which has some point of similarity with Boston.

For example, I know that Cambridge, England, has one major university and a population of about 110,000. I also know that there are at least four major universities – Harvard, MIT, Suffolk and North-Eastern – in the Boston area.

So maybe we could estimate the population of Boston as four times that of Cambridge, on the grounds that it has four times as many large universities. That would make our initial population estimate 440,000. This might be right, but it's a shaky kind of estimate, based as it is on just one point of similarity.

The evidence for our initial estimate is so weak that we really need to look for more reference points to back it up. It ought to be something about Boston you know to be similar to a British city.

So here's another angle: Boston is an Atlantic seaport, like Liverpool. I know the population of Liverpool – it's 440,000. In the past, much of the shipping between the USA and Britain was between Boston and Liverpool, so it's quite likely that their sizes are similar. In fact, the population estimate of 440,000 ties in roughly with our first estimate. (Actually, I cheated and looked it up; the population of Boston is nearer 600,000, so our estimate is not too bad.)

Back to counting pianos in Boston. First, we need the number of households. It's going to be 600,000 divided by 3 people per household. That makes 200,000 households. According to our earlier mental calculation, 1 in 20 households owns a piano. So that makes 10,000 pianos.

The final step is to work out how many pianos can be serviced by a single piano tuner before they need to be tuned again. Dividing that figure into the number of pianos will give us the number of tuners.

If you have ever lived with a piano, you'll know that a tuner might cross your threshold once a year. That is, once every

365 days. To make the arithmetic simple, let's say that there are 200 working days in a year. A piano tuner spends a couple of hours with the piano, and takes maybe an hour between jobs, so each piano takes 3 hours. But then there's lunch and things, so a tuner can do just two pianos a day. So, in the 200 working days between two tunings of your piano he will have tuned a total of $200 \times 2 = 400$ pianos.

We estimated that there are 10,000 pianos in Boston, so it's going to take 10,000 divided by 400 tuners to get round them all before they need retuning. That's 25 tuners.

Are you still sitting in that chair with your eyes closed?

SPEAK GEEK

MORE THAN 3,000 ELEPHANTS WERE SLAUGHTERED
JUST TO MAKE IVORY PIANO KEYS.

The ivory veneer on Victorian piano keys is about $1/16$ inch thick and approximately 8 inches long by $3/4$ inch wide. That amounts to $3/8$ cubic inches of ivory per key.

There are 58 white keys on a piano keyboard, bringing the total volume of ivory per piano up to 21 cubic inches. It's a reasonable guess that there were around 1 million pianos in Victorian Britain. The total volume of ivory needed in their construction would be 21 million cubic inches, or 12,000 cubic feet.

A typical pair of adult elephant tusks has a volume of about 3.5 cubic feet, so at least 3,400 elephants were killed for the ivory in our forefathers' pianos. What's worse, it's likely that many more were killed as not all of a tusk yields usable ivory.

7

BEAM ME UP SCOTTY

How long would it take to send your body to Mars?

Captain Picard has been sorting out a spot of bother on some planet or other, while the *Enterprise* orbits at a safe distance. A radio command is sent to the ship: 'Beam me up, Scotty.' His body fades out, then reappears, shimmering, on the starship.

It seems to happen so quickly, but how long would it really take? This is not just about the finite speed of light. It's much more subtle, and to understand it you need to know about my father's coffee flask.

It was standing on the kitchen table. It was 1962, my birthday, and my father had just returned from work. I knew that the flask contained a treat that had been promised for weeks.

The room was empty. Unable to resist, I unscrewed the aluminium cup from the top and eased out the cork stopper. Instantly, white vapour clouds poured over the top of the flask and rolled along the table top and down to the floor.

This was liquid nitrogen – probably the best birthday present I have ever received.

I desperately wanted this present because of the satellite receiving station at Goonhilly Downs, in Cornwall. The station had recently been commissioned, and everyone had watched the first, flickering, live TV pictures to arrive from America via the Telstar satellite.

Getting the signal from the ground to the satellite was not a problem. The TV pictures were beamed up using many kilowatts of radio power, focused onto the satellite by a giant parabolic dish antenna, like sunlight through a magnifying glass.

The difficulty was getting the signals back down to Earth. The satellite relied on small solar cells for its power, so it could generate only a few tens of watts of radio power – about as powerful as a car headlight. Worse still, the satellite was so small that it couldn't carry a big focusing antenna, so the radio waves it beamed back were spread out over a large part of the Earth's surface, an area called the satellite's 'footprint'. The footprint was about 1,000 km by 1,000 km. That's a million square kilometres. And that's why I wanted liquid nitrogen.

Let's say that the satellite transmits a radio power of 10 watts. That's about the same as the light of ten candles at a distance of 40,000 km. The task of the ground station is to 'see' those ten candles. The power spreads out over the one million square kilometres of the footprint, giving only ten-millionths of a millionth of a watt on each square metre of ground. That is a very small power.

Now here is the problem. The tiny power is picked up by the antenna of a radio receiver and causes an electrical current to flow in its wires and other electrical parts. An electrical current consists of electrons in motion. But even when there is no signal current flowing, the electrons are still in motion because the atoms to which the electrons are attached are themselves jiggling around. This jiggling generates a small electrical power in wires and other components, and that power, though tiny, is comparable to the radio power collected by the antenna.

And here is the nub of the problem. No amount of amplification will make the radio signal detectable if it is already mixed up with the electrical noise generated in the wires and other components in the first part of the radio receiver. Amplification will increase the noise as much as it does the signal, so the signal is forever swamped by the noise.

But the engineers at Goonhilly had a couple of tricks to deal with this problem. The first was obvious: they got themselves a massive parabolic dish antenna, which they named Arthur.

Arthur had an area of thousands of square metres and greatly increased the amount of radio power they could collect.

But there still wasn't enough signal to make it much bigger than the noise generated by random electron movement in the radio receiver. This is where the second trick came in: the engineers cooled all the wires and the electronics so that the atoms in them jiggled more slowly and generated less noise.

The motion of the atoms is caused by the temperature of the material. In fact, what we call the *temperature* of something is just a measure of the average jiggling speed of the molecules or atoms of which it's made. Cool everything down, and the speed is reduced, along with the electrical noise.

The cooling at Goonhilly was done by circulating liquid helium around the main electrical components. Helium boils at about −269 degrees Celsius, just four degrees above absolute zero – the temperature at which atoms cease to move – so it's very effective at reducing electrical noise.

Liquid helium is difficult to make. It's certainly too expensive to give to a twelve-year-old as a birthday present, and that's why my father gave me liquid nitrogen instead. Its boiling point is a mere −196 degrees Celsius, but that's still cold enough to give a useful reduction in radio noise.

I had planned an experiment to see if I could reduce the background noise in a home-made radio. But first I thought I'd see what liquid nitrogen would do to rubber. I lowered a piece of soft, orange rubber tubing into the flask. There was an eruption of boiling nitrogen, and I lifted out the tubing, now as stiff as a rod. I struck it with a hammer, and it shattered into tiny orange shards like a broken flowerpot.

The disadvantage of having older siblings now kicked in. My sister poured some of the liquid over the fruit bowl on the sideboard, causing an explosion of white vapour clouds. Then my brother grabbed the flask and poured liquid over the sleeve of his school blazer. Bubbles momentarily frothed on his arm. The flask was empty. There was no liquid nitrogen left for my experiment.

Nowadays we can easily pick up signals from satellites, we use tiny, low-power mobile phones, and we can listen to miniature space probes as they tour the surface of Mars. We do it all without immersing our receivers in liquid helium. So what has changed?

One important factor is brain power. One particular brain belonged to a mathematician-engineer called Claude Shannon, who laid the foundations of a subject called information theory. He published his key academic paper in 1948.

Shannon developed his ideas at Bell Labs and MIT. There was no mistaking him – he used a unicycle to navigate the long corridors to his office. Not content with the unicycle, he was reputedly designing a motorised pogo stick to further speed his travel. I remember seeing him as an elderly but sprightly man at a conference in America in the 1980s, and he was still working on methods of applying information theory to artificial intelligence. Sadly, he developed Alzheimer's disease and died in a nursing home in 2001.

One of Shannon's insights was this. In radio, it's possible to trade *time* for *power*. The power needed to successfully send and receive a signal in a noisy environment can be traded for the time taken for its transmission. He proved this with some elegant mathematics, but the idea is simple.

Imagine that you are standing on the pavement on one side of Oxford Street. A friend is standing on the opposite side. Traffic is moving in both directions, and it's almost impossible to cross. So you shout to your friend, 'I'll meet you later.' This is your message, but it gets so mixed up with the traffic noise that your friend has difficulty hearing you.

Maybe the first time you shout, she will hear the word 'you', perhaps because there was a momentary lull in traffic noise. You shout again, and this time she gets the words 'later' and 'meet'. You shout a third time, but she doesn't catch anything new. On the fourth repeat she hears 'I'll'.

She can now piece together the complete message without any errors: 'I'll – meet – you – later.' But the message has had to be repeated four times, so the time taken to convey the information was four times as long as if there had been no traffic noise.

If the traffic had been even noisier, you would have had to repeat the message even more times; less traffic, and you'd need to repeat it only once or twice.

It turns out that message repetition is just a special case of a process called *coding* that can be applied to a message before it's transmitted. There are better coding techniques than repetition. They work by replacing the original message symbols with new symbols which are worked out mathematically. The coded message is longer than the original message and so takes longer to send. But after such coding it can withstand the background noise generated inside the radio receiver, so the extra sending time is worth it. Your mobile phone uses this kind of coding.

Shannon and another electrical engineer, Ralph Hartley, were responsible for a scientific law called the Shannon–Hartley law that puts a number on the bottom line in message coding. The law says that, given an ideal coding scheme, the absolute maximum rate at which information can be sent in a message is limited by the strength of the signal compared with the background noise.

Everyone knows that radio waves travel at the speed of light in free space. What the Shannon–Hartley law shows is that, although the waves themselves travel very fast, the speed of the information they carry is limited by the background noise.

So, this is where we can start doing some rough design calculations for our matter transference beam. Specifically, how long would it take to send your body via a beam like the ones in *Star Trek*? Of course, such a system doesn't yet exist, but it's useful to be prepared. You ought to know how long it's going to take to go on your twenty-third-century package holiday from Earth to Mars.

There are several possibilities for beaming your body across space. The one I favour avoids invoking extra dimensions, parallel universes or as-yet undiscovered quantum mechanical effects. My system uses radio.

When you enter the deconstruction chamber at Heathrow Space Port, the structure of your body will be analysed down to the last cell. That includes the precise structure of each strand of DNA in each cell, as well as the composition of your body at the instant of deconstruction. The deconstruction will probably need to be destructive, so your Earth body will cease to physically exist at this point.

But the *information* about your body will still exist, stored temporarily on a computer in the Departure Lounge. When a channel becomes free, your *body-info* will be sent by radio and stored on another computer in the Mars Visitor Reception Area.

The most important device in the Reception Area is the

Bio-Constructor, a machine that re-synthesises your physical body based on the transmitted information. For this it uses proteins and other biological building blocks that are kept in dispensers on Mars.

We want to know how long it would take to transmit your body to Mars – or really, the time to transmit the information about your body structure, your body-info. You would probably prefer this to be done without any errors.

The first thing to determine is how much information is needed to describe your body down to the DNA level. One approach is to say that it's the amount of information contained in the human genome – a description of the entire useful DNA code in your body, and it is estimated to be about 6 gigabits. A bit is the amount of information given by saying that something is *on* or *off*. There are 8 bits in a byte, so 6 gigabits is the same as 750 Mbytes – an amount which could easily be stored on a computer. (The prefix 'giga-' , often abbreviated to 'G', stands for multiplication by a billion.)

Some people would point out that the information about our bodies is not stored solely in DNA. A lot of the information is implicit in the properties of the substances from which our bodies are built. That would reduce the information count still more.

But we should probably take the more pessimistic view, and reckon on having to send the information about every individual cell in your body in order to recreate it exactly as it was at the instant of deconstruction at Heathrow. There are between 100 billion and 1,000 billion cells in your body. Many of those cells have identical structures, but we would still need to send information about the presence or absence of each cell, along with its type. At the very least it is going to require

1 bit of information per cell, giving a total information count of up to 1,000 billion bits. That's 1,000 Gbits (125 Gbytes).

Now we are ready to use the Shannon–Hartley law to work out how long it will take to send your body-info to Mars. Here's the law written as a formula:

$$C = B \times \log_2 \left(1 + \frac{S}{N} \right)$$

The formula says that the absolute maximum amount of message information that can be sent in one second, C, is equal to the bandwidth, B, of the radio spectrum that can be used, multiplied by the logarithm of 1 plus the ratio of the signal power, S, to the unwanted noise power, N.

To use the formula, we need a value for B, the bandwidth of radio spectrum available for transmission. One hundred million hertz is the kind of bandwidth used for space probes, so we'll use that figure.

Then there's the ratio of the power of the radio signal received on Mars to the noise generated in the radio receiver. That's the S/N part of the formula. This is the same problem that was tackled by the engineers at Goonhilly in 1962. If we use very big parabolic dishes, both on Earth to transmit and on Mars to receive, and liquid helium to cool the radio, we might get the signal-to-noise ratio up to about 1.

Putting those numbers into the formula gives the maximum possible number of bits of information that can be sent each second. It works out at 100 million. So, to send your 1,000 Gbits of body-info will take 1,000 billion divided by 100 million, or 10,000 seconds.

That's a three-hour trip, which is about how long it takes to fly from Heathrow to Athens. The only things to worry about are a computer crash while you are in transmission, and making sure that your Dad brings the helium for cooling the radio on Mars.

Oh, and one other thing: the precise pattern of neural activity in your brain at the instant of your deconstruction has not been transmitted. Your body will arrive on Mars, but you won't remember why you are there, or even who you are.

SPEAK GEEK

IT TAKES ¼ SECOND FOR YOUR WORDS TO GO VIA
SATELLITE FROM BRITAIN TO NORTH AMERICA.

Communication satellites sit in geostationary orbits at a height of 35,900 km. A geostationary orbit is an orbit whose height is carefully chosen so that the satellite takes exactly one day to orbit the Earth. Because the period of the satellite's orbit exactly matches the Earth's period of rotation, the satellite stays above the same point on the Earth's surface all the time.

The journey time for the radio signal carrying your words is calculated by dividing the distance the signal travels by its speed. That's around 80,000 km (sending the signal up and then back) divided by the speed of light, 300,000 km per second – 0.26 seconds each way.

Undersea cable phone links don't suffer from this time delay. Although the speed of the signals along the cable is less than half the speed of light, the distance the signal has to travel is much shorter – only 5,000 km from Britain to North America. That's a journey time of only 0.03 seconds – not enough to disrupt the flow of conversation. And that's why most telephone conversations now go by cable.

8

HIDDEN DEATH

How much land is needed to bury the dead each year?

Newspaper headlines give the impression that something is terribly wrong, and Something Must Be Done:

FIVE DIE IN MOTORWAY CARNAGE

SUPERBUG KILLS HOSPITAL PATIENT

HUNDRED DEAD IN PLANE DISASTER

And yet every day, all around us, thousands of people are quietly dying, and there are no headlines to mark their passing.

The number of people in Britain is stabilising at around 60 million. Leaving aside emigration and immigration, that must mean that the number of people born every year is about equal to the number of people who die every year. But just how many people die every year?

You can estimate the figure quite easily from the average life-span in this country, which is roughly 75 years – and rising. Imagine for a moment that all births in Britain stopped today, that from now onwards people die off and aren't replaced.

Each year more die, until, after the time of the average lifespan of 75 years, there will be very few of today's 60 million people left. That means that about 60 million of us will die in the next 75 years. So, if ages are evenly spread, an average of 60 million divided by 75 people will die every year. That's about 800,000 per year. In reality, the number of people dying every year is quite a lot smaller, around 500,000. The difference between the simple estimate and the real number is because our lifespan is currently increasing quite rapidly. But, estimates are all about getting an answer that is *about* right, so we'll continue working with the figure of 800,000.

How does this figure compare with the number of people

killed in natural disasters, and major accidents? For example, how many people do you think lost their lives because of the accident at the Chernobyl nuclear plant? Was it 5,000? 10,000? 50,000?

The official report by the United Nations, the World Health Organization and the International Atomic Energy Agency puts the number of deaths caused directly by the accident at less than 50, but says that up to 4,000 could eventually die prematurely from exposure to radiation. The report caused outrage in the environmental movement, and contradicted popular belief that hundreds of thousands had either already died or would die.

Imagine that the accident had happened in Britain, with its population of 60 million and an estimated death rate of 800,000 people per year. Would it be possible to know whether people were continuing to die as a result of the accident? The first 50 deaths would be obvious enough: they would be the heroic firemen who took turns to shovel radioactive debris into the containment area. They would die in hospital, in the first weeks or even days after exposure to the intense radiation.

Let's take one of the wilder predictions of the Chernobyl death toll: say that 100,000 die prematurely from radiation-related diseases in the years after the accident (a figure 25 times greater than the UN estimate). Surely we'd be able to notice those additional deaths? The ages of all the people affected at the time of the accident are likely to match the overall age profile of the population, from infants to the elderly. The elderly might have only a few years of life left in which to develop a disease from the radiation; their early demise will be hard to pick out from standard mortality statistics. It will be the very young, who had potentially 75 years of life ahead of them, whose early deaths will be most noticeable.

If those early deaths are evenly distributed over, say, the next 20 years, we might expect the higher mortality rate in any year after the accident to be 100,000 divided by 20: 5,000 extra people dying each year against a backdrop of 800,000 'normal' deaths. The excess mortality then amounts to less than 1% of the average.

But the number of deaths each year will fluctuate because of other factors: a snap of cold weather, hot weather, food poisoning, an increase in national wealth, even social upheaval. So, perhaps surprisingly, it is unlikely that the death count we imagined from a British Chernobyl is going to be discernible in the basic yearly statistics: too many people are dying all the time for the signal from the accident to stand out from the noise of everyday life and death.

To detect the Chernobyl death signal, we need to get beyond the raw data. For example, if radioactive iodine were released in the accident, the incidence of thyroid cancers would be expected to increase in the years that followed. Look at the number of people with thyroid cancer before and after and you might well detect the effect of the accident. But if, as the UN says, only 4,000 people will die over the following decades, it may still be impossible to spot any correlation with the accident, even with this 'data filtering'. There are simply too many people dying of other causes for it to make a noticeable difference.

Anyhow, the bodies of those who die each year from whatever cause have to be tidied away somehow. Some will be buried, some cremated. The scale of deaths is easier to imagine in terms of graveyard plots, and it's easy to calculate.

You need to know the proportion of people who get buried as opposed to cremated. You also need to know the average area required for a burial. Think about the funerals you have been to. How many were burials and how many cremations? You'll probably find that it's about 30% burials and 70% cremations: more cremations in cities, more burials in the country. If 30% of those who die each year are buried, the number of burials is going to be about 30% of 800,000, which is 240,000 per year.

To get the total land area you need to estimate how much area each burial uses. In a typical cemetery, you'll probably find that graves are in rows. The graves are spaced about 2 metres apart, and the rows are about 4 metres apart. That means that each grave requires about $4 \times 2 = 8$ square metres.

Multiply the grave area by the number of burials and you get $8 \times 240,000 = 1,920,000$ square metres, call it 2 million square metres. It might help to imagine that area as a number of football pitches. The size of a pitch is about 100 metres long by 50 metres wide – an area of 5,000 square metres. We need the equivalent of 560 football pitches to bury our dead each year.

Each year, you'd need about one-third of a football pitch to bury all the people who die as a result of the Chernobyl accident.

SPEAK GEEK

A STATISTICIAN MIGHT NOT NOTICE A DISASTER IN WHICH 200 PEOPLE DIE.

In Britain about 36,000 people die in each of the summer months, and the monthly tally rises to 48,000 in the depths of winter. These are average figures. The actual monthly totals fluctuate around this underlying seasonal trend. The fluctuation tops out somewhere near 200 deaths and is called, rather portentously, the 'de-trended standard deviation'.

For this reason, a disastrous accident causing 200 additional deaths in one month would not necessarily show up in monthly mortality figures. A statistician scanning the figures wouldn't know if the blip was natural fluctuation, or a disaster.

9

IT'S FOR YOU

What is the most surprising message you will ever receive?

There's a tiny hamlet called Gallanach on the west coast of Scotland. Built into its rocky shore is the station for the transatlantic cable that is Britain's telecommunications link to the New World.

I visited this place in short trousers with my father in the early 1960s. He always arranged a 'special visit' for us children during our summer holiday. I thought it fantastic, but my brothers and sisters groaned when each trip was announced.

At Gallanach, the cables duck under the concrete floor at the start of their underwater journey across the rifts and mountains of the Atlantic Ocean. Inside the station there is a faint background sizzling sound from the electronic clocks that control the digital traffic. Once outside, the cables snake beneath the sand before plunging under the sea towards America.

One of the most famous cables was known as TAT-1. During the Cold War it carried the Hot Line between Moscow and Washington, and more than once our future depended on its reliable operation. The cable became operational in 1956, although the Hot Line agreement between the USA and the USSR was not signed until 1963, after the Cuban Missile Crisis had taken the world to the brink of nuclear war.

Phone messages between Harold Macmillan and JFK during the crisis would have been routed via the Gallanach cable. But what exactly is a message? We can think of a message as a communication that conveys new information.

Imagine that Kennedy had rung Macmillan and said that he was about to launch his missiles. That's undeniably a message, and it would be news for London. There would be warnings on the radio, sirens sounding, white-faced adults and crying children crowding into shelters.

Just how much information would there be in that worst of messages from Kennedy? Information is measured in units called bits, just as the length of a piece of string can be measured in inches.

Put yourself in Macmillan's position as Prime Minister. In a time of international crisis, he would be getting regular calls from the White House. One night the call arrives, and JFK says, 'This is it. I'm going to launch a pre-emptive strike against the Soviets.'

Now, if Macmillan had already been certain that this was going to happen, Kennedy's message would provide no new information. By our definition, therefore, the information in that message can be said to contain zero bits. Messages that say something you already know are worthless.

But what if Macmillan had been thinking that the odds of the launch were evens? Kennedy's message makes it certain. It wipes out the 50% uncertainty that existed in Macmillan's mind. A message that doubles belief in the chances of a particular event contains, by definition, 1 bit of information.

On the other hand, what if Macmillan really doesn't expect the launch? If asked, he might have put the chances at 1 in 1,000. With Kennedy's message, the launch suddenly became a certainty.

The number of bits of information in a message is defined by a mathematical formula:

$$\text{Number of bits} = \log_2\left(\frac{1}{\text{Message probability}}\right)$$

But you can get the same result without using logarithms: you simply count how many times you would have to double the original probability to get to 100% certainty. Each doubling corresponds to an additional bit of information in the message.

For example, you might start by believing that the probability of an event is 0.25, or 25%. You would have to double that twice to get to 100%. So by this measure, a message that changes your belief from 25% to 100% has 2 bits of information, and in this strange world it is quite possible to have partial doublings that correspond to a fraction of 1 bit of information.

In the case of the nuclear strike, before he got the message from JFK, Macmillan believed the chances of a missile strike to be 1 in 1,000. Doubling once brings it to 2 in 1,000, doubling again makes it 4 in 1,000, and doubling yet again gets it up to 8 in 1,000. In fact, after doubling fewer than ten times, the chances have reached 1 in 1: it's going to happen, there is no remaining doubt. In the context of Macmillan's initial near-disbelief, the message would contain just under 10 bits of information.

Rather surprisingly, the useful information in a message is not an immutable absolute: it depends on what the recipient believed about the likelihood of the contents of the message before receiving it. The amount of information in the message depends on how much it changes the recipient's beliefs.

Most of the messages we receive don't have that kind of hard edge of belief change and consequent action. Receiving a message from your aunt telling you that she has a cold may modify your actions a little – you might decide not to go round to her house for a few days. But basically, it's a weak message.

Those kinds of message don't obviously change the prior beliefs of the recipient and don't have the same kind of quantifiable information value. But their content can be quantified, by simply counting how many symbols are needed to convey the message. A symbol can be anything – a letter, a word, a number. You might get a message from a friend and count one thousand words in it: it's a 1,000-word message.

The average English word contains approximately 4.5 alphabetic characters. If you include the spaces between words, the average length is 5.5 characters. So a 1,000-word message is also a 5,500-character message.

Quantifying the information in letters, texts, emails and phone conversations in terms of the number of words or characters they contain is quite admissible, but it's more usual to convert the characters to sequences of binary symbols – 1 or 0 – and then count the binary symbols. Binary symbols are used so pervasively in computer systems that it makes sense to use the same units everywhere.

For example, in the ASCII code used to represent characters in computers, the letter A is represented by a string of eight binary symbols: 01000001, and a space between words is represented by the string 00100000. Each of those 1's and 0's counts as one bit. An average word contains 5.5 characters, so getting a 1,000-word letter through the post is the same as getting a $1,000 \times 5.5 \times 8 = 42,000$-bit letter. There are 8 bits in a byte, so your friend's letter stored on a computer would take up $44,000/8 = 5.5$ kilobytes of disk space.

Using the term 'bit' to describe both the binary symbol and the unit of information is a little confusing, but it is done because the two are very closely related. In fact, a binary symbol often (but not always) carries 1 bit of information.

Information in sounds and pictures is also counted in bits. Standard phone connections send 64,000 bits of information each second. The music stored on a CD has about 750 kilobits for each second of playing time, and digital TV uses around 1 million bits per second.

A one-hour digital TV transmission therefore consists of 3,600 million bits. The average English word needs about 44 bits to code it, so the bits used for the TV show could have been used to transmit 3,600 million/44 = 81.8 million words of text. A typical book contains 50,000 words, so 81.8 million words is the equivalent of over 1,700 books – a small library.

Let's return to those 'real' messages – ones that change your life, make you take an action or change your beliefs. What is the most surprising message you have ever received?

Surprise and the information content of a message are nearly synonymous. Surprise is what we feel when our existing expectations are radically changed. The information content of a message is a measure of how much our belief in the probability of an event is changed.

This is probably the most surprising message you are ever going to receive. You've just gone to the kitchen to make a cup of coffee, and there, standing by the fridge, is the Grim Reaper. 'I'm still young,' you object. But he just crooks his bony finger and beckons you to follow.

This really is quite surprising. Averaged over all ages, your chance of dying today is between 1 in 25,000 and 1 in 40,000. If you're under forty, it's nearer 1 in 3 million.

So numerically, just how surprising is the Reaper's message? It's equal to the number of bits of information in the message. And that is the number of times you would need to double the probability of your death before you went to the kitchen and got the message that made the probability 100%. If you're young (well, younger than forty), you'll find that 21.5 doublings take you from the 1 in 3 million chance of your death to the final certainty of 1 in 1.

So that's it: there are 21.5 bits of information in the Grim Reaper's message of your untimely death.

Here's a less surprising example: 'You haven't won the lottery.'

The chances of winning the top National Lottery prize are about 1 in 14 million, or a probability of 0.0000714%. If that is the probability of winning, the probability of not winning must be 100% − 0.0000714% = 99.9999286%. The fraction of a 'doubling' to get this number up to 100% is best worked out on a computer; it comes out at 0.0000001 bits.

That's a very unsurprising result, reflecting the very unsurprising message that you haven't won. It chimes with the saying of one of my politically correct and dry mathematician colleagues:

'The lottery is a tax on the statistically challenged.'

SPEAK GEEK

A CD CAN STORE THE TEXT OF 150 BIBLES.

If you take a Bible, count the number of words on a typical line, count the number of lines on a page, flip to the last page to find the number of pages, then multiply all those numbers together, you'll arrive at an estimate of the number of words in the Good Book. It should come out somewhere near 800,000.

Including spaces, the average English word is 5.5 characters long. That gives the Bible a character count of $5.5 \times 800,000 = 4,400,000$. A computer uses 8 bits = 1 byte of binary code for each character, so a text file of the Bible would take up 4.4 MB of storage. That's less than 1% of the space available on a CD.

A lot of Truth on a very small disc.

10
WAR CHEST

How much tax has to be collected to fight a war?

Drive north-west out of Glasgow for about an hour, and you will come to Gare Loch. The scenery is ruggedly idyllic, sunlight sparkles on the water, and ... there are hundreds of metres of high wire fencing.

This is the Faslane submarine base, home of Britain's nuclear deterrent. A short distance back from the shore, among trees and scrub, you'll find a semi-permanent camp of protesters. They're a mixed bunch: the young and rootless, the old and awkward, but also down-to-earth workaday citizens.

Of course, there are many others who don't join peace camps but still believe that possession of nuclear weapons is morally wrong. Confronted with a democratically elected government that uses taxation to fund weapons and wars, some people have tried to withhold a proportion of due tax in protest. The problems with this approach are manifest. You might not stop at withholding tax used for military purposes: childless couples might refuse to pay for education; the man who has 'never had a day's illness in my life' might claim a rebate from National Insurance contributions.

And anyway, just how much money is raised by taxation, and how much of it goes to the big-spending departments of government – defence, education and social security?

The basic rate of income tax is currently 22%. You might start by working out how much extra money is collected by the government for each 1% increase in the tax. After that you can look at what the money would buy in public sector activities.

First, we need to know the number of working people in Britain and their average taxable income. Multiplying those two numbers together, and then by the tax percentage, gives an idea of the amount of money that goes to the Exchequer.

There are 60 million people in Britain, but how many of them are working and paying tax? Let's say that the working population is between twenty and sixty years old: a span of 40 years. A simple estimate of the fraction of the whole population who are of working age is 40 years divided by the average lifespan of, say, 75 years. That's roughly half the population, or 30 million people.

Bear in mind that at any time about 10% of the population are unemployed, perhaps another 10% are ill, and half the population are women, and they will probably take time out during their working lives to rear children. We could realistically reduce our estimate of the working population by 30% to account for these factors, which gives an actual working population of 20 million.

This number is quite startling. It tells us that the population of 60 million is financially supported by just 20 million people.

Back to tax revenue. The average income of a working person is only about £20,000 – women still earn much less than men, which brings the average down. After deducting tax allowances, we're left with about £15,000 of taxable income. A 1% increase in the basic rate of tax would give the Exchequer one-hundredth of £15,000 for each person. That's £150 per person, per year. The total revenue would be the number of taxpayers multiplied by £150. That's £3 billion. With £3 billion you could order one item from the following public expenditure menu:

Make benefit payments to 200,000, single-parent families for one year

~

Build one nuclear power station that will generate 1% of Britain's electricity for forty years

~

Fund one-sixth of a submarine nuclear deterrent system for twenty years

~

Build and maintain a large teaching hospital for three years

So, how much tax does the government need to raise to fund a war? The Iraq War started on 19 March 2003. By the end of 2006 it was estimated to have cost Britain up to £5 billion – around £2 billion for each year. It can be argued that some of that cost would have been incurred without fighting a war – our soldiers would still have been somewhere – but the operational and ordnance costs would have been much lower. Assuming, in round numbers, that 10,000 soldiers are involved in the war and that each is paid an average salary of £20,000, the standing cost would be just £0.2 billion. Multiply by two to account for the usual employee overheads and it still only amounts to £0.4 billion – a small fraction of the annual war cost of £2 billion.

Our estimate showed that a 1% rise in income tax yields about £3 billion each year, so about 0.5% to 1% of personal income tax revenue has to be put in the War Chest each year to keep the war running. It's about the same level of expenditure as all the benefits paid to single-parent families.

But, the real cost, the real benefits and the real negative

consequences are incalculable. History is made only once; it's impossible to compare our long-term security, the oil price and the freedom of Iraqis with what would have been without a war. This is not a good place for Geeks to work, even the nice ones.

SPEAK GEEK

ON AVERAGE, EACH POUND IN THE BRITISH
ECONOMY IS USED EVERY 18 DAYS.

National economies have a fixed amount of money in circulation. The amount is controlled by the government by minting coins, printing notes and issuing bonds. It's called the Money Supply.

The money circulates, and the sum of all the country's transactions – shopping, buying a ship, hiring a vet etc. – in a year is the Gross Domestic Product (GDP). The Money Supply and speed of money circulation are linked by an approximate formula:

$$\text{Money speed} \times \text{Money Supply} = \text{GDP}$$

In rough figures, Britain's GDP is £1,000 billion, and the amount of ready cash in the economy – the so-called Narrow Money Supply – is £50 billion. Using the formula gives a money speed of 20, which means that each pound is used 20 times per year – once every 18 days.

11
WHEN THE WIND BLOWS

Is a storm more powerful than an atomic bomb?

Katrina, Debby, Grace: they sound so gentle and wholesome. Not at all like destructive hurricanes hurtling through the Caribbean.

Wind is air on the move. Anything that is moving has energy because of its motion, and air is no exception. The energy is usually called kinetic energy, but 'motion energy' or 'speed energy' would be just as good.

The energy in a hurricane is often compared to the energy of an atom bomb. Could that really be true? Answering the question is easier if the setting is moved to a more familiar location 5,000 km to the north-east, and instead of a hurricane we just work out the energy in a major cyclone coming up to Britain from the Atlantic.

If you are over thirty or so, you'll remember the Great Storm of October 1987. The wind speed in that storm peaked at around 115 mph (185 kph). It cut through forests like a scythe. Oaks three hundred years old were toppled like skittles. Pine and spruce trees that resisted being uprooted had their trunks snapped like matchsticks. All in all, over 15 million trees were brought to the ground by the most severe storm to strike Britain in three centuries. The scene in the forest near my Suffolk home resembled a First World War battlefield.

An interesting property of wind is that the force it exerts on an object in its path is proportional to the square of its speed. If you've ever stuck your head out of the side window of a car travelling at 60 mph (100 kph) – definitely not recommended – you'll have an idea of the force of a 100 kph wind. The force of the strongest gusts in the Great Storm would be getting on for four times as great because their speed was nearly twice that. They would have tumbled you down the street like a rag doll.

But, was the energy of the storm really on the same scale as the energy of an atomic bomb?

When a really vigorous cyclone such as the Great Storm is centred over Britain, a weather map shows the winds revolving anticlockwise in a great wheel whose axle is somewhere over England. The wind funnels along the English Channel, rips up the North Sea and turns to the west somewhere north of Scotland. Out in the Atlantic, the wind veers to the south before turning back in towards the Channel approaches for its next circuit.

The whole system might be moving north-east at around 35 kph, but the speed of the circulation around the centre of the storm can reach 80 kph. It's the circulating air that carries most of the storm's energy.

High Speed circulating Wind

DIRECTION OF STORM TRAVEL

To estimate the energy of a storm, we need to understand a few basic facts about energy, power and moving objects. Any moving mass – a car, a rock, air – contains speed energy. This energy can always be changed into heat energy or transferred to some other object by reducing the object's speed. For example, the heat generated in the brakes by bringing a car to a stop from 110 kph would be sufficient to boil a full kettle of water. All that energy was originally tied up in the speed of the car.

When a storm topples chimneys, lifts roofs or whips up giant waves on the sea, it does so by giving up some of its speed energy to move the object, and the wind speed is slightly reduced. On a walk through woodland you'll notice the stillness of the air among the trees, even on a windy day. Trees are very good at absorbing wind energy, which is why they are vulnerable.

The speed energy in a moving mass is given by this formula:

$$\text{Speed energy} = \frac{1}{2} \times \text{Mass} \times \text{Speed}^2$$

To use the formula, decide on the mass and speed of the moving object. Let's take a cricket ball, which weighs 0.16 kg, bowled at 30 metres per second. Just put the numbers into the formula:

$$\text{Speed energy} = \frac{1}{2} \times 0.16 \times 30 \times 30 \text{ joules} = 72 \text{ joules}$$

That's the energy that would keep a 72-watt bulb burning for 1 second.

The joule is a unit of energy named after James Prescott Joule, a famous English scientist who worked on energy. He was so interested in the conversion of speed energy to heat that on his honeymoon in Switzerland in 1847 he spent much of the time measuring the change in temperature of water after it had fallen down a waterfall.

We can work out the number of joules of energy in a storm in the same way as we worked out the energy of the moving cricket ball. Think of the circulating air in the storm as a car wheel with the mass of the air held in the tyre. For a real storm the wheel is perhaps 1,000 km in diameter, and the tyre is 100 km deep and as thick as the storm is high – say 2 km. The total circumference of the tyre is about three times the wheel's diameter, or 3,000 km, so the volume of air in the tyre is 3,000 × 100 × 2 = 600,000 cubic kilometres.

Air is surprisingly heavy: 1 cubic metre, the size of one of those dumpy bags of sand you see at the builder's merchant, weighs 1.25 kg. That's more than the weight of a 1-litre carton of orange juice. One cubic kilometre of air weighs around 1.25 billion kilograms, and in our storm it's all moving at 80 kph – about 20 metres per second. Putting the numbers into the formula gives:

Speed energy = $\frac{1}{2}$ × 600,000 × 1,250,000,000 × 20 × 20 joules

That works out to 150 million billion joules, or 150 million gigajoules.

Is that more or less than the energy from the atomic bomb dropped on Hiroshima? That bomb was rated at 22 kilotons – tiny compared with a nuclear bomb. A nuclear bomb can be up to 50 megatons. That's 2,000 times the power of the Hiroshima bomb.

The number of kilotons is how much TNT would have to

be used in a conventional bomb to get the same explosive power. The energy of 1 tonne of TNT is reckoned to release about 4 gigajoules when it explodes. So, the 22-kiloton Hiroshima bomb had an explosive energy of 88,000 gigajoules.

A big winter storm with winds of 80 kph going all the way round Britain has a total speed energy of 150 million gigajoules: about 24,000 times the energy of an atomic bomb, or about the same energy as a big nuclear bomb.

But the energy of the storm is spread throughout a volume of atmosphere of up to 600,000 cubic kilometres. The atomic bomb that exploded over Hiroshima released its energy into a volume over the city of no more than a few hundred cubic km, perhaps one-thousandth of the storm's volume. So the energy per cubic kilometre released by the bomb would be similar to the energy per cubic kilometre of the storm.

But much more importantly, the energy from the bomb was released in a fraction of a second, whereas the storm's energy is dissipated over several hours – a time factor of 10,000 or more. That would make the power of the bomb experienced by the people of Hiroshima 10,000 times greater than the storm. If you stood outside during the Great Storm of 1987, you'll know that the wind's force was dangerous and frightening. The power of the Hiroshima bomb would have been tens of thousands of times greater. Storms cause damage, but they don't melt concrete or vaporise steel.

SPEAK GEEK

THE SPEED OF THE GASES IN A FART CAN REACH
20 MPH.

The average speed of gas escaping from an opening can be
calculated from the volume flow rate of the gas and the size
of the opening. The volume flow rate is measured in litres per
second and can be estimated from the volume of gas and the
time in which it takes to be expelled.

Volumes are easily measured using an inverted glass tumbler
when you are in the bath. Times are measured by counting
seconds. Divide measured volume by time to get the flow rate.
Then guess the cross-sectional area of the opening. Assuming
a volume of 0.1 litres, an escape time of 1 second, and an
opening area of 0.125 square centimetres; the average escape
speed works out at 8 metres per second, or about 20 mph.
Fortunately for everyone else, the speed drops rapidly as the
gas moves away from the source.

If that sounds impressive, it's nothing compared with the
speed at which tiny droplets are expelled when you sneeze.
They can travel at 157 mph – faster than the fastest scientifically
timed tennis serve.

12
YOU'RE RUBBISH

How many skips could you fill with a year's worth of British rubbish?

Here's an ideal opportunity for honing geeky behaviour. All you need to do is hang out on your street and count rubbish bags or wheelie bins on collection day. On the other hand, if you're a bit shy and don't want to get yourself a reputation, you can do this calculation based on how much rubbish you personally put out for collection each week. But it will be more accurate if you lurk on the street.

You'll probably find that there are one or two bags for each house. Houses with just a couple of people might put out just one bag. Houses with a family of four might put out three bags. You need to work out the average number of bags per person per week. Then multiply that number first by the country's population and then by 52 to get the total number of bags of rubbish thrown out in Britain each year. Then we'll do some calculation on the volume of a bag to see how many would fit into a skip.

I'm talking here about the standard type of skip that seems to be parked on just about every British street: the ones that get filled with building detritus as we periodically replace one set of kitchen units with a shinier model, or demolish an internal wall to knock two rooms into one.

Your discreet observation of rubbish will probably give you a figure of half a bag per person per week. The population of Britain is about 60 million, so that's $60 \times \frac{1}{2} = 30$ million bags being put out each week. To get the number of bags per year, just multiply 'bags per week' by 'weeks per year'. That gives us $30 \times 52 = 1,560$ million bags per year – call it 1,500 million bags per year.

How much do they weigh? Try lifting a few bags and estimating the weight. It might be helpful to think in terms of the weight of 1-litre cartons of milk or juice. These weigh about a kilogram. Does the rubbish bag weigh more or less than a shopping bag containing ten cartons of juice? Alternatively, put a rubbish bag on your bathroom scales.

You'll find that a typical bag of rubbish weighs about 10 kg. So the annual weight of British household rubbish is about 1,500 million multiplied by 10 kilograms. That's 15,000 million kilograms. There are 1,000 kilograms in a tonne, so divide this figure by 1,000 to find the weight in tonnes – and 15 million tonnes is what we get.

So how many skips can we fill with all this stuff? We need a figure for the volume of the skip and the volume of rubbish in each bag. We'll assume that the rubbish is compacted before being thrown into skips.

The types of skip you can hire vary in size from 'mini-skips' with a capacity of a couple of cubic metres to the big ones whose capacity is around 6 cubic metres. We'll go with medium-sized skips with a capacity of 4 cubic metres. (Incidentally, 'skip' is the word in Scandinavian languages for 'ship', and skips bear a striking similarity to ships. Did the Vikings bring skips to Britain?)

As for the volume of the rubbish, you can get an idea by jumping up and down on a bag of rubbish to compact it thoroughly. I suggest you restrict this experiment to your own rubbish bags and conduct it round the back of your house to avoid getting an ASBO. I've tried this and found that, with enough jumping, the rubbish will squash down to about 50 cm by 40 cm by 10 cm. Changing these measurements to metres and multiplying them all together gives the volume as $0.5 \times 0.4 \times 0.1 = 0.02$ cubic metres.

That's the compacted volume of each of the 1,500 million bags put out each year. The total is 1,500 million $\times$ 0.02 = 30 million cubic metres.

How much rubbish will fit into one skip of capacity 4 cubic metres? Just divide the total rubbish volume of 30 million by 4 to get the answer: 7.5 million. Each skip is about 4 metres long, so if they were parked along your street bow-to-stern, they would form a line 4×7.5 million = 30 million metres long. That's 30,000 km: the line would start in your street in Britain, float in a chain down the Atlantic, round through the Southern Ocean, across Australia, up the Pacific and finish on dry land somewhere on the west coast of America.

SPEAK GEEK

THE AVERAGE BRITON GENERATES FIVE TIMES
THEIR WEIGHT IN RUBBISH EACH YEAR.

Different countries have different rubbish habits. Britons
create an average of 1.04 kg per day, in line with most other
West European countries.

But predictably, rubbish and wealth go hand in hand. US
citizens create almost twice the European average, whereas
Russians throw away only a fifth as much.

The anomaly is China. According to one source, the country
has only one-fifth of the UK's per capita GDP, but each
Chinese person creates just as much rubbish as a Briton.

13
NAPPY
CRAP

How many nappies are used each year?

It was 2 o'clock in the morning. A single, bright operating light shone above the delivery table, leaving the outer edges of the room in shadow. On one side of the table a student nurse attended to my wife as she started to give birth. On the other side, half in shadow, sat the midwife proper. She wore latex surgical gloves, and she was knitting.

As our child was born the midwife gave directions to the student nurse by pointing with her knitting needle before returning to the job in hand – some socks? A hat? I never did find out. Suddenly there was a new child, new life, and the sound of crying.

In Britain a safe birth is almost taken for granted; the UK's infant mortality rate is about 5 per 1,000 (infant mortality is defined as deaths of children under a year old). Move to Angola, and it's estimated that 198 babies die out of every 1,000 born. Even within Europe there are big differences: it's 5 for every 1,000 in the UK, but less than half that in Sweden. The difference is important, especially for the unfortunate families.

Mortality rates vary across the world for many reasons. The commonest is dehydration, but nursing practice, social conditions, climate and uneven wealth distribution all play a part as well. The association between mortality and poverty is stark, though not necessarily causal. The table shows the figures for a few key countries.

Strangely for a rich country like Britain, our preoccupation has shifted from infant mortality to dealing with the environmental consequences of successful infant life. Babies need nappies. Nappies need disposing of. There is far more talk in the media about the environmental consequences of disposable nappies than why 3,000 babies die in their first year.

Country	Mortality, per 1,000 live births	GDP, £ billion	Population, million	Per capita product, £
Angola	198	15	12	1,200
Russia	15	380	141	2,650
Poland	7	150	38	3,900
UK	5	1,110	60	18,000
Sweden	2.43	350	9.12	19,000

But are nappies really an environmental problem? To answer this question you need to know something about the number of babies born every year and how many nappies they get through. One way of getting a figure for the annual number of births is to think about people you know. More specifically, estimate the number of women of childbearing age, and the fraction of those who bear a child each year.

First, estimate the number of women who might bear children. Those women are likely to be between twenty and thirty-five years old – an age span of 15 years. On average women live till about the age of eighty, so we might assume that 15/80 of females in Britain are potential mothers. The British population of 60 million is made up of almost equal numbers of males and females, so there are 30 million females, and the number of them who are of childbearing age is 30 million multiplied by 15/80. That's about 6 million women.

How many of these 6 million women get pregnant and give birth each year? Some will never have children, and some will bear a child almost every year. Look around at the women you know who have finished having children. How many did they have on average? These days, most will have two, a smaller number will have one, and an even smaller number will have three or four. The average is probably about two over a 15-year childbearing age span. So, in any year each of the 6 million women averages about 2 divided by 15 children. That's about an eighth of a child per year. The number of children born each year is going to be 6 million × $\frac{1}{8}$ – about 750,000 children per year.

So how many nappies? If you've had children you'll know that the number of nappy changes varies a lot from day to day and from child to child. A rough figure might be three disposable nappies per day, and that goes on . . . and on . . . for about two years. It means that on any day in Britain, all children who were born over the previous two years will use an average of three nappies. That's 2 × 750,000 × 3 = 4.5 million nappies a day.

The total annual nappy count is 365 × 4.5 million, or about 1.6 billion nappies. Each nappy weighs about half a kilogram when wet. (If you really want to, you can check by putting a used nappy on your kitchen scales.) The total annual weight of soiled nappies is 1.6 billion multiplied by half a kilogram: 800,000 tonnes going into the nation's rubbish bins.

But returning to the real issue: with 750,000 babies born each year and a mortality rate of 5 per 1,000, there will be 750,000 × 5/1,000 = 3,750 infant deaths per year in Britain. And, although the actual number is a bit lower than this estimate, it's still more than all the people killed on the roads each year, in a rich, highly developed country.

So, enough of that nappy crap. Let's focus on the children.

SPEAK GEEK

THE INFANT MORTALITY RATE IN CENTRAL
BIRMINGHAM IS EIGHT TIMES GREATER THAN IN
MID-SURREY.

Overall in Britain there are about 5 deaths out of every 1,000 live births.

But there are massive variations by geographic area. For example, in 2004 the rates in the centre of big midland and northern cities were up to 12.4 per 1,000, whereas in southern rural areas the rate was as low as 1.5 per 1,000.

What's going on? Ask the man on the street and you'll get a lot of popular wisdom: unemployment is higher in northern cities than in the South; people living in the North are not so wealthy as in the South, dietary habits are poorer in the North; there are large immigrant communities in midland and northern towns; and the health budget is insufficient. The list of reasons offered is endless.

But many of these explanations do not stand up to close scrutiny. For example, there are many very poor people living in rural areas of the South of England, and the per capita expenditure on health is actually greater in the North of England than in the South.

To really understand the causes of the disparity in infant mortality requires a careful mathematical analysis that 'factors out' each hypothesised cause. Are you up to the challenge?

14
THE CUNNING
FOX

What are the best words to use in a personal ad?

Have you ever used a dating agency or a dating website? My niece took me aside to explain speed-dating.

'There are about thirty men and women in the room,' she said. 'The women sit down and then the men come round and try to chat you up for three minutes.' But here's a strange thing: 'Most of the women look pretty average – not ugly, not beautiful. But the men are much more physically varied than in normal life. They range from Mr Adonis to Mr Ugly.'

At the end of a session the players fill out a card, putting a tick against potential matches. Comparing notes with her friends, my niece noticed another gender difference. Women are selective: there are just two or three marks on their cards. Men, on the other hand, play out their biological imperative on paper. They mark more than half the women as potential matches.

A similar ritual is performed every day in the personal ad pages of magazines and newspapers. How about this for a direct, no-nonsense approach?

Cunning old fox seeks sly young vixen for fun nights out and maybe more.

I probably need some advice about this, but if you were looking for a partner, how would you sum up your personality in less than a dozen words?

Looking through these ads reveals many of the words and phrases that people use to describe themselves and who they 'wltm'. You'll find 'fun-loving', 'vivacious' and 'must have good sense of humour'. But what do these lonely hearts really mean?

What exactly are you after when you say that your ideal

partner should have a good sense of humour? Do you mean wittiness, or are you really looking for someone who will tolerate your insensitivity? Or maybe you are just so disorganised that you know it will take the patience of a saint to live with you even for one day.

Take the oft-occurring self-description of 'vivacious'. Dictionaries define it as 'brisk' or 'lively'. Those words could easily be applied to both men and women, but when did you last see a personal ad in which a man described himself as vivacious? It works both ways: not many women describe themselves as 'intelligent and solvent'.

There are lots of men who are vivacious in the sense of being brisk and lively, and there are most likely just as many intelligent and solvent women. It's just that, when applied to personality, we all know that there is much more behind those words than their literal meanings.

The phrases 'vivacious young woman' and 'intelligent, solvent man' instantly conjure up images of complete multidimensional people who surely have many other desirable personality attributes beyond the terse descriptions.

Try a word-association experiment on yourself. You might come up with something like this.

Solvent
professional
intelligent
tall
responsible
wealthy
dignified
mature

Vivacious
feminine
funny
attractive
slim
talkative
petite
fit

It looks as though personality can be boiled down to just a few well-chosen words, each carrying a complex picture of associations that we infer from our experience of life.

With this in mind, you might advise the old fox to add a sprinkling of key words to his ad, like this:

> *Intelligent, solvent, bushy-tailed old fox seeks vivacious young vixen with good sense of humour for fun nights out and maybe more.*

This probably needs a bit more work before the ad is sent off. In fact, the old fox would be well advised to look into a rigorous mathematical method of quantifying his own and his desired mate's personality. He'll then be able to pen the perfect personal ad.

Here's how.

Mathematical techniques have been used to quantify personality for decades. The search for the meaning of personality is rooted in academic research: psychologists wanted to know whether we all had the same basic ingredients in our personalities. A parallel and still debated question was the extent to which personality is governed by genes and by upbringing.

In the late 1940s, the psychologist Raymond Cattell developed a mathematical model of personality. He identified 16 basic ingredients, such as liveliness, emotional stability and warmth. Each component has a strength that is given a numerical value.

Cattell arrived at his list of personality ingredients, or *factors*, via a mathematical technique called factor analysis in which the factors are inferred from a statistical analysis of responses to a questionnaire. Each of the questions probes different aspects of personality and is assigned a score.

It's assumed that each person's score on a particular type of question will be governed by the strength of each of the 16 factors in their personality. For example, the score on a question about how much you like parties might be:

$$(2 \times \text{warmth}) + (8 \times \text{liveliness}) + (1 \times \text{emotional stability})$$

The score on another question about how upset you would feel if a friend fell ill might be:

$$(6 \times \text{sensitivity}) + (1 \times \text{anxiety}) - (5 \times \text{emotional stability})$$

In general, the score on a particular question will be a combination of all 16 factors, but often only three or four will have any significant strength, so the others can be ignored.

If a lot of people fill in a lot of questionnaires, it's possible to check how well the selected factors describe the subjects' personalities. The analysis doesn't actually tell which factors to use, just how well they work. Once a good set of factors has been found, the scores from the questionnaires can be used to make a personality profile.

This is how I think of myself. You may care to introduce yourself in the same way, maybe print it on your business card:

warmth	5	anxiety	4
rationality	8	conservativeness	4
conformity	3	self-reliance	8
boldness	6	tension	7
sensitivity	7	assertiveness	8
trustingness	5	perfectionism	6
abstractedness	3	apprehensiveness	4
privateness	7	emotional stability	6

Doing the maths gives us a numerical description of personality, but we want to advise the old fox on the best *words* with which to express his personality in his ad. We don't have the opportunity to subject him to a personality factor analysis, and even if we did, we'd still need to find the right words to express the result. A list of numbers in a personal ad is not going to impress. We need words, and to attract a mate the words must be just right.

A typical ad from a man to a woman looks like this:

> *Happy philosophical gentleman, 6 ft, passionate and romantic, into art, sport and the countryside, seeks attractive, young-at-heart slim woman who knows what she wants.*

And from a woman to a man:

> *Attractive, slim, blonde, cosmopolitan, professional woman, 35, seeks witty, wise and wonderful man to share life with.*

A good tip for the old fox would be to use some of the words chosen by other people to describe himself and his desired mate.

Taking the simplest approach, perhaps he should stick to the most commonly used words in *all* the adverts. Here they are: the top forty, statistically analysed from one hundred adverts in a national newspaper:

attractive	gsoh	kind
professional	sexy	hair
intelligent	lover	gentleman
slim	fit	generous
fun	feminine	food
tall	young	elegant
caring	wine	Christian
travel	sincere	stunning
friendship	sensitive	strong
educated	genuine	solvent
successful	blonde	shapely
share	beautiful	sensual
loving	arts	
honest	romance	

Here's a possible ad from the old fox that uses words randomly selected from the top forty:

Attractive, intelligent, professional fox seeks tall, slim, caring, educated vixen for friendship, fun nights out, and to share loving and lasting romance.

It sounds over the top, but the statistics show that this is how most personal ads read. Of course, these words are the most frequently used by both 'men seeking women' and 'women seeking men'. Maybe things would be different if we looked at these two categories separately.

Here's the top twenty of words that men use to sell themselves to women:

attractive	wine	honest
intelligent	tall	gsoh
professional	successful	arts
fun	loving	sincere
travel	lover	share
slim	caring	kind
friendship	sexy	

The first three are the same as in the top forty – it really does seem that the most important thing for dating is to be 'attractive', 'intelligent' and 'professional'. But look further down the list and you'll see that men are likely to introduce words such as 'wine', 'sexy' and 'lover'.

Look at how women appeal to men:

attractive	blonde	shapely
professional	sensitive	friendship
intelligent	gentleman	cultured
slim	fun	young
tall	feminine	stunning
educated	travel	solvent
caring	share	

There they are again: 'attractive', 'professional', 'intelligent'. But after the big three, twenty-first-century post-feminist woman typically describes herself in terms of physical attributes: 'slim', 'blonde', 'shapely' and 'stunning'.

The statistics show the bottom line – men and women want

the same main course. The difference is in the side dishes: men would like wine, whereas women go for a bit of care and sensitivity – and solvency.

So far, we've just looked at simple frequencies of words in the adverts. But mathematics can be used to do much more. For example, in terms of the advert words, is it possible to pick out distinct types of men or women?

This is a task for something called latent semantic analysis (LSA), a method involving a lot of algebra which obscures its grounding in common sense. The idea is that underneath all the apparently highly varied word descriptions there are just a small number of basic concepts.

Look at the top twenty chart for men. The words 'intelligent', 'professional' and 'tall' are all connected in most people's minds. If I tell you that someone is professional, the chances are that you'll also assume that he/she is intelligent and (as studies show) tall. And you'll probably also assume that someone to be a he.

Similarly, nobody has any problem in mentally associating the words 'slim', 'blonde', 'feminine', 'shapely' and 'stunning'. What might look at first to be a complex picture with five different attributes is really little more than a one-dimensional description: all those words point to the same mental image.

In fact, the number of broad underlying ideas that we express with language is very much less than the number of words available to describe them. Some research suggests that only about three hundred concepts underpin everything we say, hear and read.

Latent semantic analysis is used to identify the basic concepts used in a collection of texts. Each concept is defined as a combination of words, and each word is assigned a particular numerical strength. Usually, it's found that a concept will have three or four words with high strength and thousands of other words will be insignificant. The meaning of a particular text – in our case a personal ad – can be succinctly expressed as another weighted combination of just two or three of the basic concepts.

A simplified analysis of dating ads shows that there are two very strong concepts in play. The strengths of the top eight words in each are shown here for the most important concept:

Word	Strength
beneficial	0.59
mutual	0.59
generous	0.31
gentleman	0.30
beautiful	0.29
caring	0.04
fun	0.04
sexy	0.04

And for the next most important concept:

Word	Strength
affection	0.26
mature	0.26
adult	0.26
caring	0.25
intelligent	0.22
fun	0.22
professional	0.16
friendship	0.15

It's easy to make a 'map' of the adverts using these two concepts. The value of the most important underlying concept is plotted on the vertical axis, and the second most important on the horizontal axis. Each of the asterisks (man) and circles (woman) on the map represents where a particular dating advert falls in the concept space.

You can see that there are clear clusters of data points. The fox really needs to stick to words from one of these clusters, rather than using words that would place him in an unoccupied part of the concept space. For example, he could use the

words from ads that fall somewhere around the middle of the
big cluster on the left of the map. Here they are:

arts
attractive
compatible
friendship
intelligent
loving
professional
sociable
tall

We could combine these words like this:

*Tall, attractive professional fox, 49, intelligent, sociable
and arts-loving, wltm compatible vixen for friendship
and relationship.*

SPEAK GEEK

YOUR PERSONALITY CAN BE DESCRIBED BY JUST FIVE NUMBERS.

Simple models of personality combine many factors, such as *warmth*, *rationality*, *sensitivity*, *boldness* and *assertiveness*. But these factors are all related. For example, if you score highly on *warmth*, you are likely to have a similarly high score on *sensitivity*. Likewise, if you score low on *boldness*, you may well get few marks for your *assertiveness*.

This is an inefficient way of describing a personality and gives the impression that variations in personality are much greater than they really are. What is needed is a smaller set of carefully chosen factors which are completely independent of one another.

Fortunately there is a mathematical method called principal component analysis which can do exactly that. It can be used to work out sets of abstract personality descriptors that are perfectly independent of one another. Each new descriptor is a subtle combination of the original factors, such as *warmth* and *rationality*, and it turns out that only five of the new descriptors are needed to accurately describe a personality, rather than the 16 originally proposed by Raymond Cattell.

So, our personalities have only five real components. Instead of sweet, funny and expansive, with a touch of anxiety, a bit of tension, very apprehensive, warm yet rational, your personality could be described as [1, 3, 1, 6, 4]. What a catch!

15
FLY WHEELS

How many flies would it take to pull a car?

One of the key technologies for personal transport is a way of storing energy in something that doesn't weigh very much. A gallon of petrol will keep your car going for around an hour, with the engine developing an average power of, say, 20 kW. The petrol effectively stores 20 kilowatt-hours of energy in each gallon.

Petrol is an amazingly compact way of carrying a lot of energy. A gallon of it weighs only about 3.5 kg. A figure of merit for petrol as a portable fuel might be the amount of useful energy for each kilogram of fuel – its *energy density*. On this count petrol is rated at nearly 5 kilowatt-hours per kilogram (kWh/kg).

I once made an electric bike powered by a pair of lead-acid car batteries carried in panniers. Between them they weighed about 20 kg and could deliver a total energy of only 2 kWh. That's a puny energy density of 0.1 kWh/kg, just one-fiftieth as effective as petrol.

This, of course, is the curse of electric cars. There is no cheap battery technology that allows anywhere near as much energy to be stored per kilogram of battery as per kilogram of petrol.

Pedaling the bike, plus its 20 kg of batteries, back from work on yet another day when the charge had run out, I made a calculation: 1 kg of sugar digested by a human would give more muscle energy than both batteries combined.

The energy in a 1 kg bag of sugar is almost 5 kWh. Muscles can convert that energy into mechanical work with an efficiency of up to 20%. That's an output energy density of 1 kWh/kg – not as good as petrol, but, amazingly, in the same ballpark.

A muscle engine in a car would be a winner: the emissions would be just carbon dioxide, water and maybe a bit of wind, depending on the food given to the engine. And this may not be just the science-fiction dream that it sounds. There is a lot of academic research into so-called molecular motors, the power source of our muscles.

The molecular motors inside your body are built from a particular kind of protein molecule whose shape can be distorted, rather like scrunching up a piece of rubber in your fist. The scrunching is done by a chemical called adenosine triphosphate (ATP), which is supplied to the muscle in the bloodstream. ATP is itself synthesised by the sugars from your food.

At a signal from your brain, the chemical fist of the ATP is 'unclenched' – the molecule rearranges its shape into a more relaxed form, providing mechanical force as it goes. There are billions of these molecules acting together, making the muscle pull. Synthetic proteins which act in a similar way might provide the basis of future power plants.

Flying insects can do even better. They use similar molecular motors to make the wing beat repetitively with a very high energy efficiency, up to 40%. But could harnessing the power of houseflies be the way forward for car travel? If so, how many would you need to pull your car along at, say, a respectable 40 mph?

First you need an idea of how much power a single fly can generate. A very rough estimate of 1 fly-power can be made if you know the fly's weight and how long it takes to rise a certain distance into the air after taking off from a tabletop.

When the fly takes off, it uses energy to lift its weight. Some of that energy is used to accelerate its body and lift it into the air, and some will go into heating its body and warming the air. But as the fly rises, it also increases its energy in another way. This newly acquired energy is called *potential energy*. The higher the fly flies, the more potential energy it gains.

The energy is 'potential' because it doesn't come into play until something happens. If the fly stopped moving its wings, its potential energy would change into speed energy as it fell back down and would finally be released as a small amount of heat energy when it crash-landed on the table.

It's a fair approximation to say that the energy expended by the fly in rising to its cruising height is the same as the potential energy it will have once it's there. (It's actually a bit more, but for the purposes of our armchair arithmetic we can let that go.) So if we can work out our fly's potential energy, we can work out the value of 1 fly-power. We just need to divide this energy by the time it takes the fly to reach cruising height, and we have its power in watts.

The formula for potential energy is this:

$$\text{Energy} = g \times \text{Mass} \times \text{Height}$$

The g here is the acceleration due to gravity: the rate at which the speed of a falling object increases. Its value is 9.81 metres per second per second. Call it 10 to make life easier.

But what about mass – what does a fly weigh? Think of something of a similar weight that you can conveniently measure. For example, suppose that a housefly weighs about the same as a grain of rice. It could be half as much, or twice as much, but it is unlikely to be either a tenth, or ten times as much. A rice grain will do nicely.

Count how many grains of rice you can scoop up with a teaspoon, and put ten teaspoonfuls of rice onto the kitchen scales. Divide the weight by 10, and then by the number of grains per teaspoonful. You now have the weight of one rice grain – and the approximate weight of a fly. It'll come out at about 50 mg. That's fifty-thousandths of a gram, which is the same as fifty-millionths of a kilogram.

Next, have a look at some flies as they take off. How long do they take to rise to their cruising height? That's a tricky one.

You can make several rough timings and average them to get a more accurate final value. But even one timing is hard, because the time is so short. One way is to hold a ticking clock against your ear as you watch the flies rise. Older clocks have a tick-tock time of 1 second, but the ticks on some modern clocks are much more rapid. Count the number of ticks per second from your clock to calibrate your 'audio timer'.

My clock does five ticks per second, and the flies rise about 1 metre in the time of one tick – that's one-fifth of a second. Now the fly-power can be calculated. So, to put into the formula above we have the values 10 (for g), 0.00005 kg and 1 metre:

$$\text{Energy} = 10 \times 0.00005 \times 1 = 0.0005 \text{ joules}$$

Power is energy per second, and our fly has taken 0.2 seconds to rise 1 metre, so its power is 0.0005 divided by 0.2. That makes 0.0001 watts – one-tenth of a milliwatt. That's enough to make an LED glow dimly in the dark. A small battery-powered torch gives out about 1 watt – that's 10,000 fly-power. Flies aren't very bright.

A typical petrol car engine running on the flat at 40 mph will be generating 20,000 watts, or 200 million fly-power. So 200 million flies, attached by silken threads to the front of your car and suitably trained, could pull it along at up to 40 mph. Whether that's a green alternative depends on the flies.

SPEAK GEEK

THE ENERGY RELEASED BY JUMPING OFF A
100-METRE CLIFF WOULD LIGHT A 100-WATT LIGHT
BULB FOR 17 MINUTES.

Climbing to the top of a cliff takes muscular effort. A lot of the effort goes into heating your body – if you're like me, you'd be hot and sweaty by the time you reach the top. But a sizeable fraction of that energy gets stored as potential energy in the mass of your body. The amount stored is proportional to your weight multiplied by the height you climbed.

If you were to jump from the clifftop, the potential energy would be converted to heat energy when you crashed into the ground. Alternatively, if you lowered yourself on a rope that turned a dynamo, it would run a 100-watt bulb for around 17 minutes. Enlightening?

16

BUS
STOP

Can you tell the weight of a bus just by looking?

In Sir Arthur Conan Doyle's tale of the 'Crooked Man', Watson is visited late in the evening by Sherlock Holmes. Their conversation goes like this:

'I see that you are professionally rather busy just now,' said he [Holmes], glancing very keenly across at me [Watson].

'Yes, I've had a busy day,' I answered. 'It may seem very foolish in your eyes,' I added, 'but really I don't know how you deduced it.'

There is some knowledge that can be deduced just by standing, quietly observing, and applying logic to your observations. Here's one such example: stand at a bus stop until a bus arrives and pulls up just in front of you.

Look closely at the tyre on the wheel in front of you, and you'll see that it has writing moulded into the rubber. It might say: 'Inflation pressure 4 bar'. A brief moment of silent deduction and you turn to the driver and say, 'I see this bus weighs just over nine tons.'

'Yes, that's correct,' says the driver. 'It may seem very foolish in your eyes,' he adds, 'but really I don't know how you deduced it.'

'Elementary,' you reply. 'Like this.'

Like how? Well, first of all, common sense tells us that the pressure in each tyre is what stops them from being flattened by the weight of the bus. A heavy bus is going to need a high tyre pressure to stop the tyre running flat, whereas a light bus will need a lower tyre pressure. There's obviously a connection between the tyre pressure and the weight of the bus. Find the connection, and you will know the weight of the bus.

The tyre on this bus is pumped up to a pressure of 4 *bars*. The bar is a unit of pressure equal to the average pressure that the air around us exerts on every object – also known as atmospheric pressure. It's caused by the weight of the air above the ground, all the way up to the edge of space.

Atmospheric pressure is 14 pounds per square inch. That means that the atmosphere puts a weight of about 14 pounds on every square inch of the Earth's surface, so air is obviously quite heavy stuff.

Thankfully, we can't feel its weight because it pushes evenly all over our bodies on the outside and from the inside. You will feel it if you hold your breath next time you get into a high-speed lift. As the lift goes up, the external air pressure will decrease, but with your mouth closed your internal pressure will remain at the ground-floor value. The air will try to burst out through your lungs until you let your breath out through your mouth (the effect is noticeable but not dangerous).

Have a closer look at the bus. Each tyre is flattened where it touches the road. The flattened bit is about 12 inches long and 8 inches wide – the width of the tyre. No other part of the bus is touching the road, so logically it must be just that bit of tyre that supports the entire weight of the bus. Its area is $12 \times 8 = 96$ square inches.

The tyre is not flattened to contact more area of road than 96 square inches because of the pressure of the air inside it. In fact, the pressure on the inside of the part of the tyre touching the road must be just enough to support the fraction of the weight of the bus that is pushing down on that wheel.

'You see, it really is elementary,' you say to the driver, but he still looks puzzled.

· 'All right, I'll spell it out for you,' you tell him.

First, because you know the number of pounds per square inch equivalent to one bar, you can easily work out the inflation pressure of the tyre in pounds per square inch. The tyre is pumped up to 4 bars, so the pressure is $4 \times 14 = 56$ pounds per square inch.

Next, each square inch of the flattened area of the tyre is being pushed outwards by the air pressure inside the tyre. That's 56 pounds per square inch, so the total outward force is 56 pounds times the number of square inches of the flattened part of the tyre.

You have already estimated the flattened area to be 96 square inches, so the total outward force through the rubber in contact with the road is 96×56 pounds. That makes . . . well, this is just an estimate, so make the arithmetic easy by calling it 100×50 pounds. That's 5,000 pounds weight.

Finally, look at the rest of the bus. It has two wheels at the front, and a pair of wheels at the back – four wheels in all (not unusual). You have estimated that each wheel carries 5,000 pounds weight. So the total weight of the bus must be about $4 \times 5,000$, or 20,000 pounds.

There are 2,240 pounds in an imperial ton, so the weight of the bus in tons is the number of times that 2,240 divides into 20,000: about 9 tons.

'Excellent!' cries the driver.

'Elementary,' you say. 'It is one of those instances where the reasoner can produce an effect which seems remarkable to his neighbour, because the latter has missed the one little point which is the basis of the deduction.'

Revelling in your cleverness, you look up to board the bus, but it's gone.

'Smug git,' mutters the driver as he pulls away.

SPEAK GEEK

AN 8-INCH DIAMETER RUBBER SUCKER COULD LIFT FOUR MEN.

At construction sites you can sometimes see rubber suckers with handles being used to lift heavy plate glass windows. They stay stuck to the glass by the pressure of the atmosphere. The air between the glass and the sucker is squeezed out upon contact, so there is no air pressure pushing back on the glass side of the sucker. But the back side continues to be pushed towards the glass by the pressure of the atmosphere, so there is an overall force pushing the sucker onto the glass.

A big 8-inch sucker has an area of about 50 square inches, and each square inch is pushed by atmospheric pressure with a force of 14 pounds. That's a total force of 900 pounds, the same as four heavy men. Attach glass plates to the heads of the men and you could lift them all with one sucker. You could then hang them as a chandelier in your front room.

17
STIRRING AIRS

How much power does a wind turbine generate?

Where do you stand in the Great Energy Debate? Do you know how much electricity is consumed in Britain each day, or how much comes from coal, renewables or nuclear generation?

If you think that we should dramatically reduce our energy consumption, are you willing to forgo some of the comforts of modern life? Maybe you believe that we can get our energy from solar, wind or tidal sources – but will you put up with an erratic, weather-dependent supply and a disfigured landscape? You might be in favour of more nuclear power generation, but in whose backyard are you planning to bury the radioactive waste?

I have a confession. When I was a young man I proudly displayed one of those NUCLEAR POWER – NO THANKS stickers on the back of my car; I think I liked the artwork. I was genuinely concerned about the problem of waste disposal, but there was a strong element of going along with the herd.

In truth, back then I didn't know too much about what makes life go round, either physically or socially. But of course, without working things out for yourself you can't really engage with debates on the big issues as an independent and informed participant.

For example, do you know how much energy is generated by one of those wind turbines that now dot the landscape? And even if you do, do you know how much it contributes to our national energy requirement or how its output compares with the power from a conventional power station? As it happens, calculating an upper limit on the amount of power available from the wind is easier than you might think.

The job of a wind turbine is to extract energy from the moving mass of air that is the wind. The turbine slows the air, taking some of the air's speed energy, which it converts into electricity.

To make things simple, imagine the turbine placed inside a tube of width, height and length 1 metre, and place the turbine at the back end of the tube, with the opposite end pointing into the wind.

Because the air is moving through the tube, it has speed energy. The amount of this energy is given by the formula we've met before:

$$\text{Speed energy} = \tfrac{1}{2} \times \text{Mass} \times \text{Speed}^2$$

At any instant the tube contains a 'plug' of air 1 metre long by 1 metre wide by 1 metre high. The volume of this plug is $1 \times 1 \times 1 = 1$ cubic metre. Now, 1 cubic metre of air has a weight (really, we should call it mass so as not to annoy any physicists) of about 1 kg.

On British hillsides the wind speed might be 8 metres per

second. Putting these numbers into the formula gives the speed energy of the 1 kg plug of air as it enters the tube as:

$$\text{Speed energy in} = \tfrac{1}{2} \times 1 \times 8 \times 8 = 32 \text{ joules}$$

As the wind blows, the plug of air passes through the turbine at the back of the tube, and is slowed down. We can't have the speed brought down to zero because that would stop the flow of air. Slowing down by 50% is a reasonable amount. That means that the plug of air comes out of the turbine at just 4 metres per second, so its speed energy when it comes out of the turbine is:

$$\text{Speed energy out} = \tfrac{1}{2} \times 1 \times 4 \times 4 = 8 \text{ joules}$$

The difference between the speed energy of the plug going into the turbine and when it comes out is $32 - 8 = 24$ joules. That energy went somewhere, and if the design is good, most of it will be making the turbine turn the electrical generator. So we've got 24 joules from this single plug of air in the tube.

Now, that plug of air is continually being replaced by new air because the wind is blowing steadily at 8 metres per second. That means that eight new plugs of air go through the tube every second, because each plug is 1 metre long. So the turbine is converting $8 \times 24 = 192$ joules of wind energy every second.

Joules per second is called power and is measured in watts. The turbine is generating 192 watts – enough to light a couple of light bulbs, or run a TV or computer.

It's a long way from powering a whole city, but that's because we made the turbine small: the area of wind caught by the turbine was just 1 metre wide by 1 metre high. If you doubled the area of the turbine you would get twice the power because the mass of air passing through the turbine would be doubled.

We need an estimate of the area swept by the blades of the kind of turbine seen on hills in Britain. Multiply the result of 192 watts by this area, and it will give an estimate of the power generated by a real turbine. The area swept by real turbine blades is a circle, but to make the calculation simple you could approximate it to a square whose sides are equal to the diameter of the turbine blades. (If you are a stickler for accuracy, you can use the formula for the area of a circle: Area $= \frac{1}{4} \times \pi \times \text{Diameter}^2$.)

You can estimate the length of the blades when you next see a turbine. One way is to estimate the number of man-heights, house-heights or car-lengths that fit along the blade. This only really works if a man, house, or car is conveniently standing next to the turbine.

Still, you'll probably find that the kind of turbine you see in groups on hillsides has a blade length of about 10 man-heights – say 20 metres. That makes the diameter 40 metres, and using our approximation of the square area swept by the blades we get $40 \times 40 = 1,600$ square metres.

You can now work out the power generated by a real turbine: it is $1,600 \times 192$ watts. To make life easier, approximate that as $1,500 \times 200$. The answer is then $1,500 \times 200 = 300,000$ watts, or 0.3 MW.

So now you know how much power to expect from a really

big wind turbine in a moderate breeze. To join the environmental debate you need to know how many turbines there are in a wind farm, how many farms there are in the country, how often there is no wind – and more. It's also worth knowing that Sizewell B – Britain's largest nuclear power station – can generate 1,200 MW.

That's the same as the estimated power from 4,000 wind turbines – when the wind is actually blowing.

SPEAK GEEK

TO GENERATE ALL OF BRITAIN'S ELECTRICAL POWER
FROM THE WIND WOULD REQUIRE MORE THAN
100,000 WIND TURBINES.

A wind turbine can generate anything between 100 kW and
1,500 kW of power, depending on its size and the wind speed.

The average amount of power required to keep plugged-in
Britain running is 40 million kW – the same as around 13
million electric kettles. So if we make the optimistic assumption that each turbine can deliver an average power of 400 kW,
a total of about 100,000 turbines would be needed.

But if the wind didn't blow, there wouldn't be any power,
or any tea.

18
DREAM FLIGHT

What size wings does an angel need to fly?

Years ago I lived in an austere block of grey concrete flats in Helsinki. Each week we received a copy of the *Guardian*, printed on very thin tissue-like airmail paper. One dull winter Sunday afternoon, we cut our store of *Guardians* into shapes like orange-peel segments, and glued them together to make a balloon just over a metre across. More crumpled *Guardians* were hung from a string beneath an opening in the bottom of the balloon.

Outside, the crumpled newspaper was ignited, and the balloon filled with hot air and ascended slowly up the side of the flats. Then the balloon itself caught fire. Its ascent accelerated until at 30 metres above the ground it became a fireball. There was a smattering of applause from Finns who watched the balloon rise past their kitchen windows, then Sunday afternoon resumed its grey trajectory.

You've most likely experienced a flying dream. Soaring through the air, unfettered from gravity, seeing earthbound life from a new and detached perspective. Most people have experienced flying as a passenger on an aircraft, but the idea of taking to the skies ourselves seems deeply seated in our psyches and predates any kind of mechanical flight.

Witches ride brooms. Icarus strapped on wings of wax and crash-landed in the sea due to unforeseen technological failure. Modern-day fairy tales have Harry Potter riding a carbon-fibre broomstick, and children cross continents during the night in the secure embrace of flying snowmen.

And: angels manage to fly around with the aid of small white feathery wings.

The first real human flight was made by hot-air balloon in 1783. The 'aerostatic globe', designed and built by the Montgolfier brothers, carried a doctor and a military officer about 8 km across Paris. In its day, the technology attracted enormous attention from both the public and the military. Generals immediately saw the potential of flight for both surveillance and offensive action.

The development of hot-air balloon technology in the eighteenth century closely parallels the development of space technology in the twentieth. Both technologies were rapidly driven forward by military needs, and both were first tested on animals.

The Montgolfier brothers' first passengers were a sheep, a duck and a chicken. All survived the flight. The Soviets launched a dog named Laika into space in 1957. Laika didn't make it – no provision had been made to get her back to Earth.

A hot-air balloon generates lift because the air inside the balloon is less dense than the air around it. It ascends into the air for the same reason that a bubble in water rises to the surface. Ignoring the weight of the balloon's material and its payload, all that is needed for the balloon to lift off is for the weight of whatever is inside the balloon to be less than the weight of the surrounding air it displaces. The balloon floats in the air just as a ship floats in the water.

Hot air weighs less than cold air. The weight of 1 cubic metre of air at a temperature of 10 degrees Celsius is about 1.25 kg. If you managed to heat the air in a 1 cubic metre balloon to 30 degrees, its weight would fall to about 1.17 kg. The difference of 0.08 kg, or 80 grams, is how much a hot-air balloon of volume 1 cubic metre could lift. Clearly, it will take a balloon of many cubic metres to lift a man.

Say a balloon needs to lift a load of 200 kg – that's one man plus his weight again in balloon material. The number of cubic metres needed to get it off the ground is 200 divided by 0.08, which is 2,500. If you were able to paste a cubical balloon together, its sides would each have to be about 13.5 metres long. That's a cube whose sides are each a bit longer than a London bus (length, 10 metres).

But witches and children in fairy tales don't travel sedately by balloon: Mr Potter, for example, swoops and soars at fabulous speed on his Nimbus 2000. But ordinary people need wings. And wings for a man tipping the scales at 100 kg need to be big – much bigger than the tiny wings on a quidditch ball or the white fluffy things that angels in pictures have.

You don't need to be an aeronautical engineer to work out some basic facts about wings. You can use pure logic. For example, the area of a wing is going to affect the amount of lift it can generate. A large wing will give a lot of lift, and a small wing only a little. Common sense tells us that the lift will be proportional to the wing's area.

Then there is the density of the air through which the wing moves. The lift happens because of a complicated force reaction between the wing and the air. If there were no air at all there would be no lift. But if air were very heavy stuff, the forces between a wing and the air would be big. Overall, it is likely that wing lift is proportional to the density of air.

And then there is the speed. Speed is essential. No speed, no lift; stationary aircraft on the aprons at Heathrow stay glued to the ground. A fixed-wing aircraft has to accelerate down a runway until its speed is sufficient to produce enough lift for take-off. For a plane such as a Boeing 737, that might be about 150 mph.

But, does a 737 doing 75 mph down the runway get half the lift of the same plane doing 150 mph? The answer is no –

it gets a quarter of the lift. This is quite a general rule: the lift given by a wing is proportional to the square of the speed.

The usual way of explaining how speed affects lift is by considering the different distances that the air has to travel over the top and bottom surfaces of the wing. Wings are usually made so that the top surface is hump-shaped while the bottom surface is more or less straight, so that the distance that air has to travel over the top of the wing is longer than for the air passing beneath. Air passing over the top of the wing has to travel faster and expand to compensate for the greater distance. The air expansion reduces the air pressure above the wing and results in an upwards force on the wing. A mathematical analysis of the speed energy in the upper and lower flows of air shows that the pressure difference and the hence the lift force are proportional to the square of the speed.

Real wings can behave quite differently from the mathematical prediction, and common-sense logic is enough to get an intuitive understanding of the relation between speed and lift. Imagine that the wing is a slightly inclined flat sheet pushed forward through the air. Air particles collide with the under-side of the sheet, and because each particle has a bit of mass, each collision gives a little upward push to the sheet. This isn't a full picture of the way wings work, but it does describe one of the significant effects that give lift to real wings.

The upward force from each of the air particles is pro-portional to the speed at which they strike the sheet. The faster the sheet is pushed, the harder the particles strike its underside, and the bigger the lift. But that's not the whole story. The greater the speed, the more frequently particles strike the sheet. So speed affects the lift force twice over: the lift force on a wing is proportional to its speed squared.

This armchair reasoning has shown that the lift is pro-portional to the density of the air, to the wing's area and to the speed squared. Here, then, is a formula for lift:

$$\text{Lift} = \tfrac{1}{2} \times k \times \text{Area} \times \text{Density} \times \text{Speed}^2$$

You'll see that I've craftily added $\tfrac{1}{2} \times k$ in front. Don't be alarmed. The k is what is called the *coefficient of lift*, and it wraps up all the detail about how well the wing is working. The $\tfrac{1}{2}$ comes from a proper mathematical analysis of a wing.

You can test your formula against a familiar plane like a Boeing 737. I generally do this when the plane is accelerating along the runway – I want to reassure myself that it will take off and not plunge through the traffic on the M4 at Heathrow . . .

The value of k ranges from 0.1 to 1, and being cautious, I assume a value of 0.25 for the 737. Its runway speed is about 250 kph, which is roughly 50 metres per second, at the point of take-off, the total wing area is 125 square metres, and the density of air is about 1.25 kg per cubic metre. Putting these values into the formula gives:

$$\text{Lift} = \frac{1}{2} \times 0.25 \times 125 \times 1.25 \times 50 \times 50 = 48{,}828 \text{ kg}$$

And, thank goodness: the predicted lift of 48,828 kg is greater than the 45,000 kg weight of this plane. We're going to fly after all!

Let's get back to the size of wings that a man-sized angel would need in order to fly. Clearly, the wings need to generate a lift that is at least as great as the angel's weight plus any payload.

Angels, of course, have the important task of lifting souls from the dying and transporting them to the place where they are sorted for further despatch to heaven or hell. Estimating the weight of a soul is beyond the scope of this book, so we'll simply assume that the angel has to lift his own weight – about 100 kg in round numbers (he's a portly angel).

We've seen that the amount of lift depends on the speed of flight. But how fast could an angel fly? To fly as fast as a 737 he would need only tiny wings, but would need enormous power to get the speed.

Let's go for a more reasonable flight speed of 35 kph (20 mph). That's the speed of a world-class sprinter in a 100-metre race – about 10 metres per second. We'll give him a coefficient of lift of 0.1, because feathery angel wings will not be very efficient. And to make the calculation easier, we'll approximate the density of air as 1 kg per cubic metre.

Plugging these numbers into our formula, we get:

$$100 = \frac{1}{2} \times 0.1 \times \text{Area} \times 1 \times 10 \times 10$$

And turning the equation round gives:

$$\text{Area} = \frac{100}{\frac{1}{2} \times 0.1 \times 1 \times 10 \times 10} = 20 \text{ square metres}$$

So each wing is 10 square metres – perhaps 10 metres long and 1 metre wide. That's a big pair of wings.

It is possible for a mortal man to power his own flight. In 1979, Brian Allen piloted the *Gossamer Albatross* across the English Channel. The plane was made from ultra-light carbon fibre and plastic components, and was powered by furiously pedalling a bicycle-like mechanism connected to a propeller.

The total weight of Brian Allen and the plane was 100 kg, the same as we assumed for our angel. The *Albatross*'s wing area was 46 square metres, more than twice the calculated value for the angel, but it was designed to fly slightly slower, at just 30 kph (18 mph), and therefore needed bigger wings.

But at what speed *should* angels fly? The wingspan of a Sunday school angel looks to be no more than 3 metres; after all, they have to fold away neatly on his back. The width of the wings is around 1 metre, giving a combined wing area of just 3 square metres. With this wing area, the angel would need to achieve a flying speed of 25 metres per second to generate enough lift.

That's 90 kilometres per hour (58 mph), a brisk yet dignified speed for the transportation of souls.

SPEAK GEEK

A CAT COULD BE LIFTED INTO THE AIR BY A CLUSTER OF 500 HELIUM-FILLED PARTY BALLOONS.

Cats, by their nature, don't fly. Helium balloons can help. The balloons float upwards because they are lighter than the surrounding air. The helium inside a balloon is very low-density, so its weight and the fabric of the balloon we can ignore. The balloon's lifting force is therefore approximately equal to the weight of air displaced by the balloon.

The volume of air displaced by the balloon is of course the volume of the balloon itself. Calculating the volume of air whose weight is equal to the cat's weight gives us the size of balloon(s) needed for feline lift-off.

Air weighs about 1.2 kg per cubic metre, so to lift a 6 kg fat cat you will need a total balloon volume of $6/1.2 = 5$ cubic metres. You could use one big balloon, or the helium-filled party balloons you can buy in gift shops. These have a volume of about 0.01 cubic metres, so you'd need $5/0.01 = 500$ of them. But do keep an eye out for the RSPCA while you're tying them to the cat.

19
PROCESSING POWER

Which is more powerful — your brain or a PC?

An old friend once showed me the effects of Parkinson's disease. He put two beer mats about nine inches apart on the table in front of us and quickly patted each mat alternately with his affected hand. After ten pats everything seemed fine. After twenty, his arm seemed to have an invisible weight attached to it. After thirty, his hand seemed to be moving shakily through invisible treacle, despite his intense determination.

He explained it to me in layman's terms. In a part of his brain there is an electrochemical representation of his intention to move his hand. But because of the disease, the supply of dopamine, the chemical messenger that forms part of the chain of signalling between his conscious intention and the motor neurons that control his arm, becomes exhausted. The communication link is broken, and will start working again only after he's rested and the supply of dopamine is replenished at the neuron synapses.

Parkinson's disease vividly demonstrates the nature of our brains and bodies. They are quite wonderfully complex, and far from being fully understood, yet they are essentially physical machines.

Mental exhaustion occurs even in alert and healthy brains. Look at the numbers below. Work along the line, adding them up in your head as quickly as possible. Adding the first two takes little effort, adding the next is OK, the fourth is a strain . . . as you go on, it feels more and more difficult to do the next addition.

173, 184, 58, 320, 141, 90, 123, 29

Whether this exhaustion is caused by the temporary depletion of neurotransmitting chemicals somewhere in your brain, or by the accumulation of an inhibiting waste product, functionality has been temporarily reduced. Incidentally, if you truly believe that you can multitask, try doing the addition exercise while listening to the radio.

Here is a crude summary of the physical make-up of your brain. It contains some 30 billion cells called neurons, which can carry signals. Each neuron takes signals from up to 30,000 other neurons, combines them, and transmits the result to other neurons. These signals between neurons are passed as electrical impulses along connection paths to other neurons. When a signal reaches the end of its connection, the electrical impulses cause the release of neurotransmitting chemicals that drift across a gap called the synaptic cleft to another neuron, where they stimulate further signals. Sometimes, the signals may end up being transmitted via the spinal cord to muscles and other organs in the body, but often they remain within the brain, presumably as an internal representation of our conscious thoughts.

Since the late nineteenth century, there have been a succession of descriptions of how the brain functions, each drawing an analogy with the latest, complex technology. First the brain was a telephone exchange, switching messages from place to place. In the mid twentieth century it became a set of regulators, like the thermostat in an oven. Twenty-five years on, the brain metamorphosed into a computer, storing data and doing calculations.

None of these pictures is wholly inaccurate, but each one is a metaphor for some facet of our mental processes. Take the computer–brain analogy. How does your brain compare with a personal computer when it comes to doing arithmetic? To do a proper comparison, we need a numerical measure of the arithmetic speed of the two systems. For computers, the basic measure of speed is the FLOPS, which stands for floating-point operations per second. A floating-point operation is the name given to one of the basic arithmetic operations done by the processor in a computer, such as multiplying or adding two numbers together.

I just wrote a program on my computer to multiply two numbers together a billion times. The computer took 10 seconds to complete the job. So, in each second it did 100 million multiplications, each requiring a floating-point operation. So my computer can do arithmetic at a speed of 100 megaflops, or MFLOPS.

And you: what is your FLOPS rate? Personally, I would be doing very well indeed if I were able to multiply two easy numbers together every second. That's 1 FLOPS, or one hundred-millionth of the speed of my laptop.

Well, Computers 1, Humans 0.

Maybe we could beat the beastly things on memory power. The memory in a computer comes in two types. Random access memory (RAM) is temporary storage space used by the computer while it is doing its processing. Anything stored in RAM is lost when the computer is turned off. The other type is the memory on the hard disk, where files are stored more or less permanently.

Here, a simple comparison between the brain and computer memory falls down because human memory is so integral to the processing part of the brain. In a computer, memory and processing are quite separate.

For this reason, we can't simply count neurons and say that they are equivalent to the transistor memory cells in a computer's RAM. If that were true, the 30 billion neurons in our brain would be equivalent to about 30 gigabits of RAM. One gigabit is 1 billion bits of information, and there are 8 bits in a byte. On that basis, our brains would have the same order of memory capacity as a PC made in 2007.

And also, human memory is much more complex than units of electronic RAM. For one thing, memory in the brain isn't localised to a few neurons. The information in memory is distributed over thousands, even millions of different neurons. Destroying modest numbers of these neurons doesn't cause memory to disappear, although the destruction of a very large number of them, for example after a stroke, does impair the overall neural function, temporarily or permanently.

Another big difference between brain and computer memory is to do with definition. You can store photos of your family or music files for your iPod in a computer. The exact pixel description or sound samples stored in those files can be used to recreate, or *remember*, the picture or music indefinitely.

Human memory is different. Most of us can't recall images and sounds in that literal way. What you store in your brain is a higher-level abstraction. You might 'remember' the waves, sand, sunbathers and rocks you saw on the beach on your holiday. But you won't be able to draw a particular wave you saw, or describe it in detail. That's impossible.

Most people don't have a literal memory, but a few have an eidetic, or photographic memory, which seems to allow them to remember the precise detail of a scene or music. At least, they can correctly answer questions down to a level of detail that suggests they have total recall. That rare ability is often linked to autism and idiot savants.

One of the best-known individuals with such power is Kim

Peek, the American on whom the movie *Rain Man* was based. We can get an idea of the potential memory capacity of the human brain by looking at some of Kim's feats.

For example, it is said that he can read a book and remember 98% of it word for word. Actually quite a large number of rather normal people can remember books word for word. My mother can quote large extracts from the Bible, and Muslim scholars can do the same with the Koran. The thing about Kim Peek is this: he can repeat the feat for the *twelve thousand* books he has read so far, and presumably for all the books he is going to read in the future. (He can also do it after a single reading – a feat most religious scholars don't attempt.)

So, how many gigabytes of memory are needed to store the contents of 12,000 books? This is an easy question to answer if you have a computer with a word processor.

Open a document and click on 'word count' to get the number of words in the document. Then save the document as plain text. Use the explorer to see how many bytes are being used to store the document. I did this for a document on my computer. The document contained 1,897 words and was stored as a file of 11 kB. That means that each word in the document needs an average of $11,000/1,897 = 5.8$ bytes per word.

Now estimate the number of words in those 12,000 books remembered by Mr Peek. A typical book has 250 pages, and each page might contain 400 words. That's 100,000 words in a book. So in 12,000 books there are $100,000 \times 12,000 = 1.2$ billion words. Kim managed to remember only 98% of the words, but that would still be near enough 1 billion.

Each word needs an average of 5.8 bytes to store it, giving a total storage requirement of 5.8 billion bytes for 1 billion words. That's 5.84 gigabytes of memory; about twice the RAM in a good PC but just a small fraction of the storage available on a hard drive. So even Kim Peek's prodigious memory is only on the same scale as computer RAM.

You might argue, quite rightly, that Kim knows lots of other things besides the contents of those 12,000 books. But even if he has a hundred archives similar to his book memory, he would still have a total memory capacity comparable only to that of a large computer hard drive.

I make the score now Computers 2, Humans 0.

Surely humans are better than computers at remembering pictures? Why not give all the benefit of doubt to humans on this one? Imagine that you have eidetic memory and that you have remembered in perfect detail every image projected onto your retina, for every blink of your eyes, ever since you were born.

How many gigabytes of memory of image data would be stored in your brain? To work this out, you need the following numbers: the rate at which you blink, your age and the number of bytes of information in each image.

I surreptitiously observed a member of my family at breakfast, and they seemed to blink about once every 5–10 seconds. That's 6 blinks per minute, 360 per hour. If a person is awake for 16 hours in the day, that'll be 5,760 blinks each day, around 2 million blinks each year.

By the time someone reaches their fiftieth birthday, they'll have blinked about 100 million times. And for the purposes of our calculation, each time they blink they'll need to capture and store the scene in front of their eyes.

If you were to take a photograph with a digital camera, a single image might have a size of 500 kB after it is compressed into the JPEG file format. So the total memory size required for the 100 million images is 100 million multiplied by 500 kB. That's 50,000 GB.

Now, 50,000 GB is a lot more memory than you have on the hard drive of your PC, although with 30–100 PCs you would have it. In a few years' time you may be able to buy a PC with more memory than this. Then you could store every image you have seen each time you blinked from your birth till your death.

That makes the score Computers 3, Humans 0.

Isn't there some way for humans to come out on top in this comparison? After all, who would you prefer as a friend – computer or human? Perhaps it's best not to answer that.

The problem is this: we keep playing the computer–human matches on the computer's home ground. Human brains haven't evolved to add and multiply symbolic representations of numbers at high speed. We designed and built computers precisely because we aren't very good at that kind of task. That's why they're called 'computers', and that's why they win hands down.

But when it comes to picking out a face in a crowd, or recognising that two pieces of music carry the same emotional message, humans perform much better than computers. So, one last heave for humans to show that we're the masters here. We need a test that humans have evolved to tackle, but which could also be programmed into a computer.

Let's take looking at an image of a face and recognising its owner. We think we are quite good at this task. Some people seem amazingly good at it. A friend of mine was given the task of showing the Queen and Prince Philip around his workplace and introducing them to a hundred or so staff and local worthies. The Queen was delayed, so my friend set off with just the Prince on the great handshaking tour.

When the Queen arrived, the Prince said that he would take her around to repeat the introductions without my friend's help. He did it without making a mistake, and remembered every name.

'How did you do that?' asked my friend incredulously when he got back. 'Ties!' replied the Prince. Instead of trying to remember a hundred human faces, he had focused on the different patterns on ties and associated them with the owner's name.

The Prince's task was unusual because it required only a short-term association. But how many faces could you reliably recognise, perhaps when shown a series of photos? For myself, I think it's five hundred to a thousand, and I can probably recall a name, relationship or some labelling context in a few seconds or less.

Computers are programmed to perform this task, known as pattern recognition, by storing a mathematical representation of each of the known faces. Presented with an unknown face, the computer compares it with each of these representations, and the name associated with the representation that gives the best match is assigned to the unknown face.

One of the commonest methods of matching the stored representations of known faces to the unknown face is to look for periodic variations in brightness across the image. The measurement is done by a mathematical algorithm called the fast Fourier transform, or FFT for short.

The FFT is the single most important mathematical algorithm used in mobile phones, digital TVs, iPods and other modern electronic devices. Just about anything that processes images or sounds has the FFT algorithm coded into its processor.

The invention of the FFT is usually credited to the computer scientists James Cooley and John Tukey in 1965. The algorithm was actually invented much earlier, but its importance wasn't recognised because there wasn't any technological application for it.

Its real inventor was Cornelius Lanczos, a mathematician and physicist who worked as an assistant to Albert Einstein in the late 1920s. Cornelius did the original work on the FFT in 1940, but never got the credit. He eventually moved to a university in Dublin where an Irish friend remembers him as a man who always wore a black beret, and assumed him to be a refugee from Franco's Spain. In fact, he was from Hungary – another country of beret-wearers.

The FFT's great virtue is, as its name implies, its speed. For example, using the FFT to compare two 1 megapixel images requires over 80 million separate multiplication and addition operations. Previous algorithms would have taken at least a million million operations – quite impractical, even with the fastest computer.

A computer using the FFT to perform the Prince's recognition task would have to do 80 million pieces of arithmetic for each of the 100 candidate faces. That would make 8 billion operations. Your PC will do about 100 MFLOPS: that's 100 million operations each second. So it'll take 8 billion divided by 100 million = 80 seconds to process each face.

If you plan to replace the royal visitor with a PC on wheels, there's going to be quite an embarrassing pause before each handshake as the computer works its way along the line of expectant worthies. Prince Philip can do it in less than

1 second and shake hands at the same time. We really are quite impressive. The final score:

Computers 3, Royalty *much more.*

SPEAK GEEK

An ADULT CAN PERFORM ABOUT 360 ARITHMETIC SUMS PER HOUR.

Many years ago, the Post Office Savings Bank used hand-operated adding machines to work out the quarterly balance and interest due on each account. The results were then checked by young women who used no more than their heads.

The Post Office required each of them to process 60 accounts per hour. The record of each account would typically have six different amounts (like £3 7s. 3d.) to add up, and then be multiplied by a percentage such as 2% to find the interest due. That's nowhere near the speed of your PC, but it's still pretty impressive.

20
SOUL
MATES

Weird coincidence: chance or design?

I have a friend who lives in a tiny Suffolk hamlet. He was brought up in Kent and moved to Suffolk for his work. His father was the captain of one of the last commercial sailing ships in Britain, and was of German descent. His mother's parents had moved to England from South Africa after his aunt had caught the plague and died in her childhood in the early 1900s. They burnt all their domestic belongings before leaving for fear of carrying the disease into their new life.

In his sixtieth year, my friend decided to trace his roots. He found that his great-grandparents were buried in a churchyard just three miles from his Suffolk home. He had always assumed them to be South African.

Raahe, a small town in Finland, was the setting for one of the strangest coincidences in recent years. One day early in 2002 there were two road accidents on the main road through the town. The accidents happened within two hours and within one mile of each other. In both accidents a lorry struck a man riding a bike. Both accidents were fatal, and both riders killed were men aged seventy-one. They were twins, each unaware of the other's movements that morning.

We humans are sensitive to apparent connections between widely separated events within our sphere of perception. Are these freak chance events, or is some strange force at work? It can be straightforward to work out whether what looks like weird coincidence is the result of mere chance or of more mysterious forces. Take those twin brothers in the north of Finland – was that chance or something else?

With this kind of question it's important to start by defining the boundaries of space and time within which you would become aware of a noteworthy event. We live in a world saturated by news coverage – in paper, broadcast and electronic form – and we would probably hear about an event like the one in Finland if it occurred anywhere in Europe or North America, and we can include much of Asia too. Let's say that's about 3 billion people within the geographical boundaries of our perception.

What do we mean by the time boundaries of the event? It's about how long something remains in our memory. You might recall a strange story several years after the event, and if the event was very dramatic you might keep the memory all your life and recount it to your children and grandchildren. At that point, it could enter folk memory and have an effective lifetime of hundreds of years.

An example of a folk memory is a story about the execution of King Charles I of England. Three and a half centuries later, people still recount the tale of the king wearing two shirts on the scaffold to prevent him shivering with cold and leading spectators to think he was shaking with fear.

Strange events like the one that happened at Raahe may stay in our memories until we die, and on average people will be halfway through their lives when they first hear the story. That would put the time boundary of the event at 40 years. We'll ignore the fact that some people are forgetful, even of the strangest things.

With the time and space boundaries of our perception fixed, it's possible to start working out the natural chances of the twin death in Finland. If the likelihood of its occurrence by pure chance works out at, say, 50%, most people would accept that no other forces were at work. If the odds turn out to be a thousand to one, many would claim the action of stranger forces.

We could start by getting a figure for the number of cyclists killed each day on the roads. In Britain there are about 3,000 road deaths per year, and 5% of those are cyclists. That's about 150 cyclists killed each year out of a population of 60 million, or about one every two days.

The road casualty rate in Britain is quite low compared with other developed countries, and has halved since 1970, even though the amount of traffic has doubled. A more realistic fatal cycle accident rate across much of the developed world would be twice the British rate: say 1 death every day for every 60 million people. Put another way, the chance of any individual dying in a cycle accident on any particular day is 60 million to 1 against. (These odds are so high because of the low number of cyclists and their low chance of being killed in an accident.)

The Finnish town of Raahe has a population of less than 30,000, so by simple scaling we would expect a cyclist to be killed only once every 2,000 days. That's about one every five years, and the odds of a cyclist dying on a particular day are 2,000 to 1 against. Very rarely, even in a community of just 30,000 people, two cyclists will be killed on the same day, but the odds are tiny: just 2,000 × 2,000 to 1. That's 4 million to 1.

The accidents in Raahe were even less likely than that. The two victims weren't just random individuals within the community – they were twin brothers. To refine those odds, we need to know the number of twins in Raahe.

In the developed world, the chance of being a twin is about 1 in 40. So the odds against someone being killed on a bike on a particular day in Raahe, *and* being a twin, are 40 times the usual odds: 80,000 to 1.

At first sight, the chance of the second twin dying is much

smaller because we are looking at the probability of the death of a particular individual; the second victim has to be the first victim's twin brother or sister. If the behaviour of the second victim is not linked in any way to the first, those odds are just 60 million to 1 – the normal probability for dying in a cycling accident.

Without any link between the deaths, the odds of first one twin and then the other twin being killed on the same day are staggeringly small. It's got by multiplying the probabilities of the two separate deaths together. That's a chance of 1 in 80,000 for the first twin, multiplied by 1 in 60,000,000 for the second. The odds are 4,800,000,000,000 to 1 against.

But, these phenomenally low probabilities are misleading. There are almost certainly factors linking these two deaths. For example, twin brothers are likely to have similar lifestyles. In particular, if one is a cyclist then it's likely that the other one is too. And once someone gets on a bike, the probability of their being killed in a cycle accident is much higher than average.

We could take an even more extreme position and suppose that twin brothers have identical habits to the point where their every action is identical. Both brothers would leave the house together, ride alongside each other and incur the same risk of death in the same accident: if one dies then so will the other, because their behaviour is identical.

In this extreme scenario the chance of both twins dying at the same time and place is the same as that of a single twin dying, because they would occupy almost the same space at the same time as each other. That brings the odds of both twins dying on the same day on the same road down from 4,800,000,000,000 to 1 to just 80,000 to 1.

The real odds of the twin death lie between the two extremes; exactly where, we don't know. One way forward is to be conservative and assume that a single factor links the brothers.

Let's start by saying that because the first brother was a cyclist, the second was bound to be a cyclist. The chance of the second brother being killed in a cycle accident is then much higher than for the average person, who is unlikely to use a bike. In fact, if we assume that about only one in ten people regularly use a bike, the odds against the second brother's death by cycling reduce by a factor of ten, from 60 million to 1 to 6 million to 1.

That brings the odds of the dual death down to a mere 480,000,000,000 to 1. But even after arguing for these lower odds for the dual death, it still looks like the Raahe event is too unlikely to happen by chance. Maybe some other force is at work after all . . .

Except that there's something we haven't yet taken into account. The remote odds apply to all the days within our memory and to all the communities within our sphere of news awareness. Let's take the memory period first.

The odds of the twin cyclists dying on the same day are 480,000,000,000 to 1, but this is where the time boundary becomes important. If you are aged eighty, you'll remember significant events that happened on any day within your eighty-year lifetime. On the other hand, if you have just been born, the number of days you remember is almost zero. On *average* people will remember events that took place on any day within half of their final lifetime. That's about 40 years, or roughly 15,000 days. The chance of a twin dying on *any one* of those days is 15,000 times greater than the chance for a single day. That brings the odds down to 32 million to 1 against.

Now, 32 million to 1 still sounds rather unlikely, but it applies to just one community with a population of 30,000. We might well remember such an event if it had taken place in any of the communities across our news domain.

The population covered by our news is about 3 billion – nearly half the world's population. Imagine the 3 billion people divided into communities of 30,000, roughly the same as the population of Raahe and its immediate neighbourhood. There would be 100,000 communities of that size. In each community, the odds of an accident happening are 32 million to 1, but there are 100,000 communities, so the chance of an accident in *any one* of them are multiplied by 100,000. That makes the odds of us hearing about such an event 320 to 1.

Those odds are still high in relation to commonplace events, and we might still be tempted to start looking for an explanation other than bad luck. But *chance* is still in with a chance. This is all to do with how we register events – our psychology.

The strange event in Raahe was the death of twin brothers in cycle accidents, but any similar story would also have made the news. For example, a husband and wife dying in separate bike accidents on the same day, next-door neighbours, father and son, mother and daughter, people with the same birthday, and so on. There is an almost endless list of newsworthy relationships that can be woven around the underlying event of two people dying in cycle accidents on the same day and in the same town.

Let's allow a hundred such relationships into our reckoning – and then call a halt. Each of those variations on the basic story has a similar probability of 1 in 320, but the probability of any one of the stories occurring is 100 times greater. The odds now? Just 3.2 to 1.

We have been quite conservative throughout this calculation, particularly over the linkage between the brothers' behaviour, and in limiting the number of story variations to just a hundred. Even so, the chance of this kind of event registering in our awareness is now close to 30%.

With a touch of maths, the weird becomes likely, and mysterious forces that intervene in human affairs are consigned to the file where the conspiracy theories and alien abduction reports are kept.

SPEAK GEEK

AT A PARTY WHERE THERE ARE 23 PEOPLE, THE
LIKELIHOOD OF ANY TWO OF THEM HAVING THE
SAME BIRTHDAY IS GREATER THAN 50%.

Line up the partygoers and ask the first person, a woman, to step forward. There are 365 days in a year, so the chance of her having the same birthday as the second person in the line is 1 in 365.

The chance that she shares a birthday with the third person is of course also 1 in 365. And so on. Each of those individual odds will add up to give the total chance of her sharing a birthday with someone in the line. If there are 23 people, the probability is $23/365 = 0.063$.

Now, repeat the calculation with the second person, but don't count the chance of sharing a birthday with the first person again. The probability this time is $22/365 = 0.060$. Continue the process and add up all the odds. With 23 people the sum will work out at 0.5005, or 50.05%.

It's a small world.

21

IDIOT CALCULUS

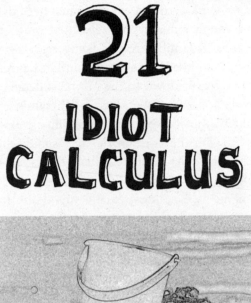

What can you work out while sitting in a deck chair?

If you've read the novel *The Curious Incident of the Dog in the Night-time* by Mark Haddon, you'll remember how the main character, a young autistic boy, tells how much he dislikes familiar objects in familiar rooms being rearranged. Change makes him feel unsettled, even afraid, until he has remembered the new arrangement down to the minutest detail.

Most people seek comfort and security in familiar surroundings. For many, the comfort lies simply in the existence of a familiar house with the same worn carpets and scuffed furniture, the electric kettle and teapot just where they always are in the kitchen.

And some of us, I confess, extend this desire for comfort through familiarity into a desire to quantify our surroundings. Take, for example, a child's sand bucket on the beach. Do you know how many grains of sand it holds? Until you know, can you really lay back in your deck chair and relax? I can't.

Imagine a grain of sand placed against the scale of a ruler marked in millimetres. The grain's diameter might be around 1 mm. But the average grain size varies from beach to beach. On a beach of coarse sand it might be 2 mm; for fine-grained sand it can be 0.5 mm. The grain size is going to make a very big difference to the number of grains in the bucket.

If the diameter of each grain were 1 mm, it would just fit inside a tiny box whose sides were 1 mm long. The volume of that box would be $1 \times 1 \times 1 = 1$ cubic millimetre. You just need to work out how many of these boxes would fit into the bucket by working out its volume. The number of little boxes will be equal to the number of grains.

Buckets are an awkward shape: a truncated cone, like a cylinder with a taper. You could do some geometry to work out the volume precisely, but an approximation will be good enough. Any error in your estimated bucket volume will be dwarfed by the uncertainty in the size of a sand grain.

Beach buckets are about 200 mm high and have a diameter of around 150 mm at the top and 100 mm at the bottom. You won't make a big error by using the average diameter, and taking the bucket to be a cylinder of diameter 125 mm and height 200 mm.

The volume of a cylinder is its height multiplied by the area of one end. That area – and the area of any circle – is the mathematical constant π multiplied by the square of the radius. But for this estimate it's good enough to do a box approximation again. Imagine that the cylinder is held snugly inside a cardboard box, like those boxes used to pack fancy whisky bottles.

The height of the box is 200 mm, and its width and depth are both the same as the diameter of the bucket, about 125 mm. The box will have a greater volume than the real bucket because it has space at the corners that isn't occupied, so you could reduce the effective width and depth of the box to 100 mm to compensate. The volume of the box (and so of the bucket) is about $200 \times 100 \times 100 = 2$ million cubic millimetres.

A bucket of volume 2 million cubic millimetres will hold 2 million of those 1 mm diameter sand grains.

What if the sand is coarser than you originally measured, say with 2 mm diameter grains? Or, what if the grain shapes are so irregular that they don't pack nice and densely like little balls or boxes?

You could make the estimate again assuming a sand grain size of 2 mm. With 2 mm grains, the volume of each little box is $2 \times 2 \times 2 = 8$ cubic millimetres. That'll bring the number of grains in the bucket down to just 2 million/8 = 0.25 million. Or maybe the grains are smaller, just 0.5 mm in diameter. That would allow 8×2 million = 16 million of them to be packed into the bucket.

Well, that feels better. The sand bucket is now fully understood and presents no threat to your peace of mind. But there are many other things that can be calculated as you laze in your deck chair. For example, as you look up you might see the vapour trail of a jet, and then spot the plane itself as a silver sliver at the trail's head. Could you work out the speed of the jet without getting up?

One way of working out its speed is to count how many seconds it takes the jet to fly from directly overhead until it disappears over the horizon. Just divide that distance by the time and get the speed. But, how far will the plane have travelled by the time it reaches the horizon?

Jets fly at a height of about 10,000 metres, and if you are not near an airport, the chances are that the jet is at cruising

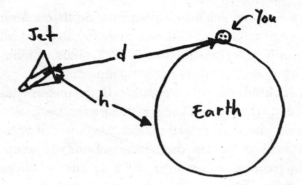

height. Working out the distance at which a plane at this height just seems to touch the horizon involves a bit of simple trigonometry and knowledge of the Earth's diameter. And there's a simple formula. The distance is equal to 3,600 times the square root of the height. Everything in this formula is in metres, and it works both ways: if your viewing height is h, then it tells the distance of the horizon at sea level:

$$d = 3,600 \sqrt{h}$$

The jet's height is 10,000 metres. The square root of 10,000 is 100. So, it will disappear over the horizon when its distance is 3,600 × 100 = 360,000 metres away. That's 360 km.

But hang on – that's too far, surely? Even if the air were clear and you had eyes like a lynx, it's going to take the jet forty minutes to go 360 km, and the tide might be lapping round your feet by then. We need another way.

An easier though less accurate plan is to hold the palm of your hand flat against the sky at arm's length. Look with one eye, and count how long it takes the plane to pass behind your hand. You'll find it's about 6 seconds. Your palm is 10 to

12 cm across, and you'll be holding it about 80 cm from your eye.

The jet is 10,000 metres away: 12,500 times as far as your hand. So it had to travel 12,500 palm-widths to appear to cross your hand. That's 12,500 × 0.1 = 1,250 metres. The plane travelled 1.25 km in 7 seconds, and there are 3,600 seconds in an hour, so its speed must be 1.25 × 3,600/6 = 750 kph.

That's two quite complicated calculations you've managed without recourse to computer, book or internet, and if you are anything like me you'll be starting to feel quite rested. But look around – there's more to work out. Like the power station behind my local beach. It hums away twenty-four hours a day, seven days a week, sending out electric current along cables suspended from giant pylons.

Electric current is a flow of atomic particles called electrons that have almost no mass. So, a question you might ask from your deck chair is what weight of electrons has been delivered from Britain's power stations today?

Strictly, the answer is none. In Britain, electricity is supplied as an alternating current of electrons. That is to say, the electric field along the wire is reversed fifty times per second (hence the description 50 Hz) – there is no net flow of electrons. (This has the same effect as a continuous one-directional flow of electrons; a so-called direct current.) The electrons move backwards and forwards a very short distance without, on average, making any overall shift in either direction. However, what you could do is count the number of electrons that just enter and just leave the power station.

Electric current is measured in amperes, or amps for short. One amp (1 A) is one unit of electric charge delivered in one second. A current of 20 A means 20 units of electric charge per second. A tiny packet of electric charge is carried by each electron, and it's the moving electrons that carry the current.

The charge carried by a single electron was first determined at the end of the nineteenth century, but its value was not accurately known until the results of experiments by Robert Millikan were published in 1913 and won him a Nobel prize. Millikan managed to attach small numbers of electrons to tiny droplets of oil falling through air. The electrostatic attraction of the droplets to electrically charged metal plates allowed him to deduce the charge on a single electron. It worked out at about 1.6e – 19 units of charge.

An electric current of 1 A is a flow rate of one unit of electric charge each second; it is carried by 1/1.6e – 19 = 6.24e18 electrons per second. A current of 1 A running for 1 hour would transfer 6.24e18 × 3,600 = 2.2e22 electrons.

A typical home uses about 10 kilowatt-hours of electricity per day. With around 23 million domestic properties in Britain, the daily domestic energy consumption will be about 23 million multiplied by 10 kWh – that is, 230 GWh each day.

Electrical power is current multiplied by voltage, so the current is calculated by dividing power by voltage, which in this country is 230 volts. That works out as an average current of 42 million amps for 24 hours every day. The number of electrons that flow in a current of 1 A for 1 hour is 2.2e22, so the total daily electron flow to the nation is 2.2e22 × 42e6 × 24, which comes out at 2.2e31 electrons.

This is a phenomenally large number of electrons. It is 22 followed by thirty zeros, so the total might be expected to weigh quite a bit. But one electron has a tiny mass, just 9.1e – 31 kg. That's a decimal point followed by thirty zeros, then the digits 9 and 1. The total mass of the electrons sent out daily from our power stations to service the electrical energy consumption of the nation for one day is 9.1e − 31 multiplied by 2. 2e31: about 20 kg. That's about the weight of a sack of potatoes, which you could lift with one hand (if you're fairly strong).

The tide has come in, with the calming predictable echo of planetary motion. It's time to leave the beach for home. I do hope there's nothing incalculable on the way.

SPEAK GEEK

A CHILD WALKING ON A STONY BEACH FEELS
ONE-THIRD OF THE FOOT PRESSURE FELT BY AN
ADULT.

The discomfort you feel walking barefoot on a shingle beach
is caused by the pressure of the sharp stones on the sole of
your foot. The average pressure over the sole of your foot can
be calculated by dividing your weight by the area of your sole.
Of course, the pressure at the point where a sharp stone sticks
into your sole is much greater than the average – that's why it
hurts – but the ratio of average to peak pressure will be roughly
constant for a given type of shingle regardless of whether you
are an adult or child. Comparing the average pressure for an
adult and child will give a measure of the relative discomfort.

Stand on a piece of paper and draw round your foot with a
pen. Now, remove said foot and, with a ruler, measure the
length and widest part of the outline. Multiply those two
numbers together to get the area of a rectangle that would just
fit round your foot. Of course, it's more than the real area of
your foot, but there's no need to worry about that. Divide that
number into your weight, in kilograms, and you'll have a
number that is proportional to the average pressure on your
foot.

Repeat the operation for a three-year-old; you'll find that
your pressure is up to three times greater. Children today have
it too soft.

22
THE GHOSTLY PRESENT

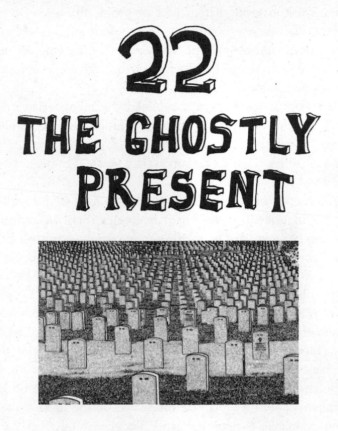

Do the dead outnumber the living?

Is the world overpopulated? Lots of people think so. Do you think Britain is overpopulated? How about your town, even your family home? Well, of course, that's different. *We* definitely have a right to be here; so do our family, our friends and 'people like us'.

Overpopulation is always caused by someone else. It's a fertile field for growing all sorts of conflict, and ostensibly is based on religion, social class and race. At its root lies the age-old competition for resources that we want to preserve for ourselves and 'people like us'.

Some ecologists define overpopulation as a state in which the population of an organism is greater than the numbers that can be supported by the resources in the organism's habitat. Grow bacteria in a Petri dish with a limited supply of nutrients, and eventually they will stop multiplying because there is nothing left to fuel further division.

In 1789, long before bacterial multiplication was understood, the political economist Thomas Malthus applied the same idea to humans in his *Essay on the Principle of Population*. He believed that, in the absence of any control, the human population in any particular locality would increase until their numbers could no longer be supported by the food available from accessible agricultural land. People would then die of starvation, leading to a catastrophic collapse in population.

The food–population system would not necessarily reach a balance: instead it might oscillate through many cycles of population expansion followed by starvation. The unstable fluctuations in population are down to time delays. It takes time for food production to be increased, just as it takes time for infants to grow into full food-consuming adults. The time delays nearly always make the system overshoot the balance point at which the available food just matches the population.

The population of Britain fluctuated around 4 million for many centuries until the Agricultural and Industrial Revolutions. By 1800 it had reached 8 million, and then the growth curve began its current climb. We're still clinging on to its steep, slippery surface at a population level 7.5 times greater than that of 200 years ago. The reasons for this 52 million increase in population are contentious and complex, but two factors stand out: fossil fuel, and technological knowledge.

Oil is still very cheap: at the time of writing (mid-2007) the wholesale price is about $70 dollars per barrel. A standard barrel contains 35 US gallons, or 159 litres, so the price works out at 22 pence per litre. That's about half what you pay for it as central heating oil, or half what you pay at the petrol pump before duty. The difference between the wholesale and retail price is down to refining, distribution and 'middle-man' costs.

Think what you can do with a litre of oil. Put it in the tank of a 200 hp tractor and you'll be able to plough non-stop for up to 15 minutes. I watch the 20-acre field behind my house being ploughed every autumn. It takes one man with a big tractor about 5 hours – that's one acre every 15 minutes.

Before the Industrial Revolution, a man and horse might plough an acre a day. Even in industrialised Britain, men were routinely using horse-drawn ploughs up until the Second World War. A relative told me how, as a young farm-hand, he pushed his horse to do more than an acre a day. But, he said ruefully, 'The horse knew when the acre was done, and wouldn't do more.' A modest tractor-drawn plough is around thirty times faster than a horse-drawn plough, for a cost of just one or two pounds-worth of diesel per acre.

Of course, just being able to perform the various processes for growing food thirty times faster doesn't necessarily provide thirty times as much food. There has to be enough land to grow the food, and it has to be fertile land. Until the Agricultural and Industrial Revolutions, soil fertility was maintained by putting animal and even human manure on the fields. Arable farming was also restricted to naturally fertile land.

Since then, artificial fertiliser has made much more land usable and productive. But fertilisers such as ammonium nitrate have to be synthesised chemically using energy from fossil fuel, or dug out of the ground and imported from far-away places, again using fossil fuel, to power the excavation machinery and for transport.

Our present system of food production may not be sustainable; the Malthusian catastrophe could still happen. The king of fossil fuels is coal. Each year on our planet we burn an amount of coal that took a million years to form. The story for oil is similar. At some point they'll run out.

The rapid increase in the world's population over the past couple of centuries, driven by the exploitation of fossil fuel, has led to speculation that right now there are more people alive than have lived before in the whole of human existence. But how true is that – are there really more of us walking the Earth than lying six feet under?

To uncover the truth, it's clear that we need two numbers. First, how many people are alive on the planet at say, twelve noon today. And second, the total number of people who have ever lived, *and died*, since the apes came down from the trees.

The Earth's human population today is about 6.6 billion. Almost none of those people were alive more than 100 years ago. What's more, they are nearly all going to die over the next 100 years. That's at least 6.6 billion deaths over the next century. That's the key to answering the question. We can count all the people who have ever died by going back through time in steps equal to a lifespan. At each date we'll add the population of the world to a count of the dead.

The counting will be easy if we assume that everyone lives for the same number of years, equal to the average lifespan of a human. We're dealing with total numbers across the planet, so there is nothing wrong in using an average number. These days we British can look forward to a lifespan of around eighty years. But for much of the world's population, life is shorter. Maybe sixty to seventy years is nearer the mark.

Going back five thousand years, we find from archaeological evidence that our own ancestors lived only into their thirties or forties. So, for the purposes of counting the dead, let's assume that the average lifespan over the whole of human history has been fifty years.

If everyone lived to the age of fifty and then died, we could reason that the entire population of fifty years ago, that's 1957, is now dead. If we knew the world's population in 1957, we could use it as the first contribution to our dead-count. That's just the start. Go back another fifty years to 1907, and suppose that all those people were dead by 1957. Add them to the dead-count and repeat the addition until, say, 50,000 years ago to get a rough figure for the dead.

You may have spotted a small problem with this approach: the number of steps of fifty years to cover 50,000 years is 1,000. Do you really want to add a thousand numbers to get the answer? To avoid anxiety on this score, I can reveal that accounting for deaths in just the last few hundred years is going to be sufficient for a good estimate of the number of people who have ever lived and died. More of that later.

To do the count, you need to know the world's population at each of those fifty-year intervals. You might be able to look it up in a book or on the internet, but it's more interesting to estimate it yourself. Here's how.

The population is growing. It grows because, on average, each set of parents has more than two children who live to have children of their own.

The British are often quoted as having, comically if not accurately, 2.4 children in a family. That would mean, on average, one person becoming 1.2 people after about the period of one generation – say twenty-five years. After another twenty-five years those 1.2 people have children. The number goes up to $1.2 \times 1.2 = 1.44$. That's an increase of 44% in the size of the breeding stock over fifty years – about 1% per year.

The interesting thing is not the actual numbers, it's the numerical process: more people means more children means more children. The result is that the population increases by a certain factor in a given time period. For example, say the population increases by 44% over fifty years. The increase in population after 200 years would be 1.44 multiplied by the number of fifty-year intervals in 200 years. That's a factor of $1.44 \times 1.44 \times 1.44 \times 1.44 = 4.3$.

The birth rate in Britain is now lower than in most other countries, and our population trend isn't representative of the rest of the world. Fifty years ago, the world's population was about half the current size of 6.6 billion, so on a planet-wide scale it appears that the population seems to roughly double every fifty years. You can use this multiplication factor to write a table of estimates for the population going back through

time in fifty-year steps. Every time you go back fifty years, just
divide the population by two:

Year	Estimated population
2007	6,600,000,000
1957	3,300,000,000
1907	1,650,000,000
1857	825,000,000
1807	412,500,000
1757	206,250,000
1707	103,125,000
1657	51,562,500
1607	25,781,250
1557	12,890,625
1507	6,445,312

The table is surprising – and *wrong*. It's telling us that 500
years ago the world's population was just 6.4 million. This
goes to show how a simple mathematical model can give completely false results if we push it too far, in this case by trying
to extrapolate a long way back into the past.

It is always worth running a sanity check on a numerical
estimate. The idea is to compare the estimate with one or two
pieces of indisputable evidence (they are often called ground
truths). For example, historical records show that in 1500 the
population of Britain was about 4 million. Today it is nearer
60 million. That's a ratio of 15 to 1. But according to our
table, populations have grown by a factor of a thousand since
1507. Conclusion: the table is wrong.

So the way in which populations grow is much more

complex than our simple idea of people breeding like dividing bacteria in a Petri dish of unlimited nutrients.

Instead, we'll have to rely on historical and archaeological records. These show that the world's population was about 5 million in 10,000 BC and grew to between 200 and 400 million by the time Jesus Christ was born. It then rattled around that figure until, between 1700 and 1800, things began to behave much more like our table, with the population doubling every fifty years. In fact, the numbers in the first five rows of the table are about right.

Well, after all that, you can now count the dead and answer the question: are there more people alive today than have ever died? The number of dead is the sum of all the population numbers at fifty-year intervals starting at 1957 and going back to, say, 10,000 BC. The population started at around 5 million at the beginning of this period and finish up at 400 million before the Agricultural and Industrial Revolutions. We could take an average population of 100 million.

It takes 240 steps of fifty years to get back to 10,000 BC. That's 240 lifespans. The number of people who have died in that period is the average population of 100 million multiplied by the number of lifespans. That's 240×100 million = 24 billion: about four times the current living population.

So, the answer to the question is that the living *don't* outnumber the dead. On the other hand, our table does show us that the number of people alive today is greater than the number of all the people who lived and died in the past couple of centuries.

SPEAK GEEK

IF PRESENT POPULATION TRENDS CONTINUE, IN
400 YEARS' TIME THERE WILL BE JUST 100 SQUARE
METRES OF LAND FOR EACH PERSON.

The total land area of the Earth is about 150 million square
kilometres. If you divided it into little plots of 10 metres by
10 metres – about the total footprint of a house on a new
estate – you'd end up with 1,500 billion of them.

The world's population is currently around 6.6 billion and
has been doubling every fifty years or so for the past two
centuries. To find how long it would take for the population
to grow to 1,500 billion, count in steps of 50 years and double
the population each time until it gets to that figure. It's around
eight steps – that's 400 years. There would then be just one
10 × 10 metre plot for each person.

It gives new meaning to 'pressing the flesh'.

23

BAD BREATH

Get a sniff of Caesar's last gasp.

In 1948 a graduate student called Clair Patterson began work at the University of Chicago on a way to determine the age of the Earth. His method relied on measuring the amount of lead and uranium in old rocks. The trouble was, he found that his rock samples were contaminated with lead – from the atmosphere.

The lead in the air came from lead tetraethyl, a compound which until recently was commonly added to petrol to make it burn more smoothly. The lead added to petrol worldwide from the 1920s onwards is still around today in the atmosphere, the oceans and the ground. It's toxic, but we have no choice but to live with the consequences of our past. It is the ultimate bad breath.

Lead tetraethyl was first used as a petrol additive in 1921, by an engineer turned chemist, Thomas Midgley. But his contribution to long-term environmental damage didn't stop there. In 1928 he developed a chlorofluorocarbon (CFC) as a non-toxic replacement for the ammonia that was used in early refrigerators. It's now accepted that CFCs are the main cause of ozone depletion in the upper atmosphere, as well as being a potent 'greenhouse gas'.

Our atmosphere is not infinite. Some of the things pumped out by various human activities break down harmlessly; others hang around for a very long time. The Earth is more or less a sealed unit: the only substances that leave the planet in significant quantities are the very lightest gases, hydrogen and helium, which tend to slowly boil off into space. Some other substances, such as compounds of lead, are eventually absorbed into rock, but only after many millennia. Meanwhile, they've taken up residence in our bodies, plants, oceans and soil.

Now, breathe in, and hold your breath.

Your lungs now contain a sample of the atmosphere-borne pollutants of our modern industrial age. But that's not all – the litre or so of air now inside you also includes small quantities of all the stuff that has been pumped out by people, plants and machines over many centuries.

Take Caesar's dying breath. You probably just inhaled a few of the molecules from the air that passed between his lips as he collapsed on the steps of the Senate and gasped 'Et tu, Brute?' You can work out just how many molecules if you know the volume of his lungs, of your lungs and of the Earth's atmosphere.

Of course, we have to assume that Caesar's last breath has been thoroughly mixed into the whole atmosphere and hasn't been permanently absorbed somewhere. The time it takes for the atmosphere to be recycled by dissolving in the oceans and then being released, or being transpired by plant life, is an interesting question in itself – but one we'll have to put to one side.

Start with the volume of air in Caesar's last breath, and the volume of air in the Earth's atmosphere. Dividing the first number by the second will give the tiny fraction of a litre of Caesar's breath that's now present in each litre of atmosphere.

Next you have to work out how many litres of air are inhaled in one breath. Multiplying that number by the concentration of Caesar's breath in the atmosphere will give the actual volume of his breath you inhale. It's going to be minuscule, but there are a huge number of molecules in even one litre of air, so even in a small volume there will still be a fair number. To get that number, just multiply the volume in litres by the number of molecules per litre. That will give the number of molecules of his dying breath you just breathed in.

It's time to put some actual numbers into this estimate, starting with the volume of one lungful of air. A very deep breath amounts to one litre, about the same as a carton of orange juice.

Then there's the volume of the Earth's atmosphere: that's the area of the Earth multiplied by the height of the atmosphere. You could look up the area of the Earth with Google, or use the formula for the area of a sphere ($4\pi r^2$) to work it out from its radius (6,370 km). It comes out at about 500 million square km.

The thickness of the atmosphere is more difficult to pin down because its density and pressure aren't the same all the way up. The air at ground level is squeezed by the weight of all the air on top of it, so it's quite dense. The air at Mount Everest's summit doesn't have so much air above it. Less weight of air on top means that it isn't squeezed as much, and therefore it's less dense.

There's a simple way of dealing with this by thinking in terms of an 'effective thickness' of the atmosphere: how high it would be if its density all the way up were the same as it is at the surface. Some undergraduate maths shows that the effective thickness of the atmosphere is the height at which the pressure of the real atmosphere has dropped to 37% of its surface value. That turns out to be about the height of Everest, which is near enough 10 km.

Multiplying the Earth's area by the effective thickness of the atmosphere gives a volume of 5 thousand million cubic kilometres. There are a million million litres in a cubic kilometre, so the volume of the Earth in litres is

Effective atmosphere thickness

Mount Everest

Earth

Air

5,000,000,000,000,000,000,000. That's 5 followed by 21 zeros, or in our shorthand notation, 5e21.

So if Caesar's last breath had a volume of 1 litre, it forms one part in 5e21 of the air we breathe. If your last breath was 1 litre, you just breathed in $1 \times 1/5e21 = 1/5e21$ litres of Caesar's last breath.

All that remains is to estimate how many molecules there are in this tiny volume. It's done using an important number that's named after the nineteenth-century Italian scientist Amedeo Avogadro.

Avogadro was a professor of physics at the University of Turin. He was also a bit of a political activist, and was removed from his post in 1823 so that he could 'spend more time on his research' – the nineteenth-century equivalent of 'spending more time with one's family'.

Avogadro realised that a weight, called the *mole*, of any substance always contains the same number of molecules. In the 1860s, the Austrian physicist Josef Loschmidt worked out what Avogadro's number was (Avogadro himself wasn't able to calculate it), and used recent discoveries about gases and molecules to show that 1 litre of any gas contains around 2.7e22 molecules.

That's the number we need. That lungful of air you breathed in contains $2.7e22 \times 1/5e21 = 2.7e22/5e21$ of the molecules that were in Julius Caesar's last gasp. That works out to be 5.4. It's an average. Some breaths will have none of these molecules and some will have lots. One thing's for sure – there's no need to worry about whether Caesar had halitosis.

Oh, I almost forgot – you can stop holding your breath now. Let it out, and think about all the people who'll be breathing it in a couple of thousand years from now.

SPEAK GEEK

AFTER TOPPING UP YOUR QUARTER-FULL CUP OF
TEA TEN TIMES, THERE WOULD BE LESS THAN
ONE-MILLIONTH OF THE ORIGINAL MILK IN THE
CUPPA.

Many people have the habit of drinking half a cup of milk tea
and then topping up with tea from the pot without adding
more milk. If you drink three-quarters of each cup before
topping up, only a quarter of the original milk content
remains. So, after ten top-ups the final milk concentration will
be reduced by a factor of $1/4 \times 1/4 \times 1/4 \times \ldots \times 1/4 = (1/4)^{10} = 1/1,048,576$. That's pretty strong tea.

A similar process of successive dilution is used in the prep-
aration of homeopathic medicines. A solution of the supposed
therapeutic ingredient is diluted, typically with 99 parts of pure
water or alcohol to one part of the original solution. The
process doesn't stop there: the dilution is repeated, this time
starting with the already diluted ingredient. The process of
successive dilution is repeated up to 15 times before the final
solution is administered to the patient.

A calculation similar to the one for working out the strength
of your cup of tea gives the concentration of the homeopathic
solution. It's $1/100 \times 1/100 \times 1/100 \times \ldots \times 1/100 = (1/100)^{15} = 1e - 30$.

But a 300 ml cup of the solution would contain a total of
only about 1e25 molecules. That means that the number of
molecules of therapeutic ingredient in the cup would be just
$1e25 \times 1e - 30 = 1e - 5$, or 1/100,000. It's not possible to have

a fraction of a molecule: what this means is that there is a chance of only 1 in 100,000 that there will be one of those molecules in the cup of solution you drink. The proponents of homeopathic remedies get round this by suggesting that the water or alcohol used in the dilution process 'remembers' the earlier presence of the active ingredient.

Maybe.

24
WAVE POWER

How many electric kettles could be powered by a wave?

'Sometimes I like to sit and think, and sometimes . . . I like to just sit.'

What do you do on the beach when you go on holiday? Do you slap on the factor 30, spread out your towel, lie back and let your mind go blank? Or, do you slap on the factor 30, spread out your towel, lie back and . . . think? Do you, for example, watch those waves slop lazily onto the sand and wonder how much power they have?

On a calm day, the waves are probably less than a metre from trough to peak. They might arrive at intervals of about 10 seconds, and be spaced 20 metres apart. Of course, the power will depend on the length of wave taken into the calculation. You might restrict it to 50 metres to your left, and 50 metres to your right. That's a 100-metre length of wave.

The calculation seems simple enough in principle. The waves are travelling at a certain speed. All moving mass has energy. If you could just work out how much energy is carried along by the waves each second, that would tell you their power.

But waves are slightly more subtle than that. If you watch waves for a few minutes, you might spot something – a piece of seaweed, perhaps – floating a few metres out from where the waves are breaking. As each wave passes, the seaweed merely rises and falls with the wave, but doesn't actually move towards the shore. The water in the waves isn't really travelling anywhere, it's just going up and down.

This is what's behind wave power: extracting energy from waves. Each time a wave passes, it lifts its own mass through an average of half the trough-to-peak height. Raising any object – an iron weight, an aircraft, an elevator cage or water in a wave – stores potential energy in the object that can be released when the object is allowed to fall. Build the right kind of mechanical system, and that released energy can be collected.

One way would be to anchor a float out at sea. It would rise and fall with the waves, and it would be easy enough to mechanically connect the rising and falling float to an electrical generator. We've met the formula for potential energy before. It's

$$\text{Energy} = g \times \text{Mass} \times \text{Height}$$

where g is the acceleration due to gravity, the rate at which the speed of a falling object increases. Taking g to be 10 metres per second per second will give us a good enough estimate. The amount of energy released by an object of mass 1 kg falling through a height of 1 metre is therefore about 10 joules. If the height were 10 metres or the mass 10 kg, the energy would be 100 joules, and so on.

To work out the energy in a single wave, you need to know the mass of the water in the wave, as well as the average height through which it falls. The fall is half the trough-to-peak wave height, which on this calm day is about 0.5 metres.

The mass of water in the wave is harder to work out because it has a strange humped shape. It could be very roughly approximated by a rectangular shape, so that the waves come along with vertical leading and falling edges.

The horizontal distance between successive waves is about 20 metres, so the block of water in the peak of the wave has a cross-sectional area of $10 \times 0.5 = 5$ square metres. Of course, real waves don't have sharp edges, so their area will be much less – say 2 square metres.

In a 100-metre long section of wave, the total volume of

water in the wave peak is therefore about $100 \times 2 = 200$ cubic metres. One cubic metre of water weighs 1,000 kg, so the mass of this water is 200,000 kg. That 200 tonnes of water in the wave falls through a height of 0.5 metres, so from the formula above, the released energy is $200,000 \times 0.5 \times 10 = 1,000,000$ joules.

The waves are coming in once every 10 seconds, so the energy per second carried by the waves is $1,000,000/10 = 100,000$ joules per second. A joule per second is another way of saying 1 watt of power. So, a 100-metre length of the waves breaking languidly on your beach carries a power of about 100 kW. If all of that energy could be extracted, it would power thirty electric kettles.

That would make a nice cup of tea, just the thing to go with your picnic lunch on the beach.

SPEAK GEEK

THE BOXING DAY TSUNAMI WAS EQUIVALENT IN ENERGY TO EXPLODING A LARGE CAR BOMB EVERY 100 METRES ALONG THE SHORE.

The Indian Ocean tsunami was up to 24 metres in height. That's about three times the height of a house.

An average 1-metre long section of the wave would have a volume of up to 300 cubic metres and mass 300 tonnes. Its energy is the mass of water multiplied by its average height multiplied by 10.

With an average height of around 8 metres, a 100-metre section of the wave has an energy of 2.3 thousand million joules; equivalent to the simultaneous explosion of a large car bomb every 100 metres along the coast.

25
MECHANICAL NEUROSIS

How much do you love your father?

This is a story about Jean. Jean isn't sure why she feels so unhappy. Maybe it's just the human condition, or maybe it's something that happened in childhood that she can't admit to. Her relationships don't last. They end with disappointment, and not so much in anger as with a sense that it can't work out. What's more, she feels that she just doesn't deserve to be happy. Her life is washed grey by guilt; she is to blame, but she doesn't know what for.

Jean loves her father. He's getting on a bit now, but she still feels like a little girl when they meet. She's always got on well with her mother, and when she was in her early teens they were inseparable.

They were a happy family, Jean thinks. But just occasionally she sensed a coldness between her parents, or heard muffled angry voices after she'd gone to bed. Why couldn't the two people she loved get on? What had she done wrong to make them hate each other?

If you were asked to put your finger on the root of Jean's problems and their solution, what would you say?

'Pull yourself together and get on with your life.'

'Your father has treated your mother like a doormat. Be realistic: he isn't adorable. Stop seeing him and look after yourself.'

'Your mother has never grown up. She used you to perpetuate her own childhood, and denied you yours. Recognise the situation and move on.'

'You can't love both your parents if they don't love each other. You'll be torn between them unless you choose one and reject the other.'

One view of Jean's problem is that her neurosis arises because her emotional beliefs aren't self-consistent. From the description of her feelings you might deduce that, on the one hand, she loves both her mother and her father, but on the other, she dislikes anyone who doesn't treat her mother or father well.

The trouble is, she realised in childhood that her father and mother actually dislike each other. So how can she love her father if he dislikes another person she loves, her mother?

In the 1960s a psychoanalyst turned computer scientist, Kenneth Colby, wrote a computer program in an attempt to model personality as a set of beliefs. He wanted to investigate how free association – that's seeing what other words come into a person's head when they hear an apparently random word – could be used as a therapy and to get to a patient's core beliefs. Of course, the program comes nowhere near describing the complexity of real people, but it does give a glimmer of insight into what might be going on in our heads at some time or other in our lives.

Understanding how Colby's program works is easy if you imagine yourself pretending to be the computer. This is a common ruse used by computer scientists; it's called a *Wizard of Oz* experiment. For example, computers are decidedly inferior to humans when it comes to recognising speech. But it is still possible to find out how people would deal with a twenty-second-century computer that had perfect command of the English language by letting subjects believe that they are talking to a computer, when really there is a human operator in another room responding to their words. The experiment allows the restrictions imposed by the human/machine interface and the computer's poor knowledge of the world to be investigated without being swamped by the effects of imperfect computer speech recognition.

This is the role I would like you to take: you are the computer-based psychoanalyst, a Wizard of Oz, acting out the algorithm encoded in Colby's program. You would probably let Jean talk to you via a computer keyboard and respond in the same way, so that she sees your replies on a screen.

According to Colby's algorithm, you would start by probing Jean with questions to uncover the depths of her feelings about her parents. You need to measure the vehemence of her denials and affirmations when you contradict or agree with her beliefs, and then use that measurement to infer the belief strengths.

This process is inherently inexact, and will have to be repeated many times before you have reliable belief scores, but eventually you should be able to determine the maxima and minima of the various belief strengths and put them all on a scale from 0 to 10.

In the real computer program, the scores would be recorded as a matrix in its memory. The scores are on a scale running from 10, which represents a very powerful belief, down to 0, which is complete ambivalence. The table might look like this:

No.	Belief description	Strength
1	I love my father	8
2	I love my mother	6
3	Mother dislikes father	3
4	Father dislikes mother	9
5	I dislike anyone who dislikes a person I love	5

The diagram shows the extent of Jean's neurosis. Just look at the strengths of the 'belief signals' coming from Jean to her mother. One signal, saying that she *dislikes* her mother, has a strength of 45, and the other signal, saying that she *loves* her mother, has a strength of 6:

After studying the diagram and seeing how much she needs help, you would be ready to start applying the therapy. The algorithm runs close to Freudian principles, or at least Freudian as interpreted by practitioners such as Melanie Klein.

You would begin by coaxing Jean into voicing her beliefs. Next you persuade her to accept their unacceptable, simultaneous, incompatible coexistence. Finally, you set in train a series of tiny, individually imperceptible changes in the strengths of her beliefs until they reach a more self-consistent value.

This might work, and a lot of people have spent a lot of time and money over the past hundred years subjecting themselves to this kind of care. Some believe that it has profoundly changed their lives. Others say it's all psychobabble.

This is not the only therapy that's amenable to computerisation. There is a program called ELIZA that is based on a non-directive psychotherapeutic approach made popular by the psychologist Carl Rogers. ('Non-directive' means that the therapist tries to coax the patient into finding their own solutions.) ELIZA is dumb. Even her creator, the computer scientist Joseph Weisenbaum, agrees that she's dumb. In fact, it was part of his mission to show how humans can make relationships, even make intimate friends, with an algorithm that simply returns statements with a question or assertion which sounds as if it has come from an empathetic human.

Here is an example of ELIZA talking to a patient:

> CLIENT: 'My boyfriend says I'm *depressed*.'
> ELIZA: 'I am sorry to hear that you are *depressed*.'
> CLIENT: 'It's true, I'm *unhappy*.'
> ELIZA: 'Will coming here help you not to be *unhappy*?'

Weisenbaum's program has a list of about forty key words, including *depressed* and *unhappy*. If the patient uses one of these words in a statement, the computer detects its presence and reuses it in a question. Most of the time, the result is what sounds like a sympathetic and wise human response, but there is absolutely no natural or artificial intelligence in the system.

The Colby system is much more sophisticated. For example, if you managed to persuade Jean to change the strengths of her belief that her parents hate each other to zero, the neurosis would disappear and she would be able to love both her father and her mother. She would have no internal conflict because she would no longer believe that her parents hate each other.

Or, you could try to get Jean to stop believing that she loves her mother (Belief No. 2). That would leave her free to dislike her mother without internal conflict, and all the other beliefs could be left at their original strengths.

That last solution might be a bit unethical – and anyway, there are other solutions that reduce Jean's internal conflict by a greater amount. One approach is to gently push her towards incremental changes in each of her belief strengths. That's much easier than trying to make her completely reject an existing belief.

But you would need to know how to make the correct adjustments by using a few of the mathematical equations encoded in the computer, and this is the single most important part of the algorithm: there has to be a numerical measure of the discord of the belief set, calculated from the strengths of the beliefs. Given this measure, it's straightforward to search for new belief values that reduce the discord or neurosis.

In an artificial intelligence experiment I once watched, a robot mouse was equipped with a light sensor and a heat sensor, and was programmed to steer away from the dark and towards heat. These were the mouse's two core beliefs. Placed in a room with radiators in the shadows around the edges and more light towards the centre, the mouse navigated its way across, doing its best to satisfy both its love of heat and its fear of the dark. It minimised its primitive neurosis by finding a compromise path to steer.

A similar method is used in the search to tackle Jean's neurosis. Here are some of the points you might suggest to minimise her internal conflict.

First, if there are no limitations on how much her beliefs can be changed, there is a perfect solution. She should believe that everyone loves everyone else with exactly the same intensity. Her belief values could be 10, 10, 10, 10 or 3, 3, 3, 3 or 0, 0, 0, 0. It doesn't matter how much love there is, as long as it's the same between all parties.

For various reasons, that solution is unlikely to be acceptable. Here's another: leave Beliefs 3 and 4, 'Mother dislikes father' and 'Father dislikes mother', with their original strengths, but persuade her to love her father and hate her mother by equal amounts. If Jean were to continue loving both parents, there would be a conflict caused by Belief 5, 'I dislike anyone who dislikes anyone I love.' Persuading Jean to hate her mother makes Belief 5 consistent with all the other beliefs. The result is perfect internal harmony.

A similar approach is followed every day on thousands of computers seeking the best compromise solutions to all sorts of problems: finding the best design of a building for which there are conflicting requirements for warmth and ventilation, light and privacy; planning the best route for petrol tankers taking fuel from depot to filling stations; or finding a layout of the micro-circuitry in a mobile phone that minimises cost while maximising performance.

So if you want to know how much you love your father, write down your set of core relationship beliefs and start optimising: minimise your neuroses.

SPEAK GEEK

ONE PERSON IN EVERY FOUR WILL SUFFER FROM A
MENTAL HEALTH PROBLEM AT SOME TIME DURING
THEIR LIFE.

You might think that illnesses such as depression are sympto-
matic of modern affluent societies, but according to the World
Health Organization, depression is the fourth most common
illness in the world and affects the poor more than the rich.

In Britain, a significant proportion of GPs' time is spent
helping patients whose problem stems from a lack of mental
rather than purely physical wellbeing. As a result, many general
practices employ counsellors and therapists to deal with the
problem, and therapies such as Cognitive Behavioural Therapy
are already available on line.

It's only a matter of time before therapy administered by a
Rogerian or Kleinian computer algorithm becomes available
through NHS Direct.

Which would you prefer, PC or Mac?

26

THE FINAL JUDGEMENT

How will your account look when you die?

'And I saw the dead, small and great, stand before God; and the books were opened: and another book was opened, which is *the book* of life: and the dead were judged out of those things which were written in the books, according to their works.'

That's the cheery Judgement Day line-up described in the Book of Revelation. So, how will the *Book of Your Life* read when it is opened? Is it possible to put some numbers to the sum total of your life?

Of course, there are many things that could be included in the *Book of Your Life*. The credit column of the ledger might list the number of small acts of kindness you performed, the number of children you successfully raised and the amount of wealth you created. The debit column will record how you were disloyal to friends, the times you allowed greed to direct your actions, and maybe your personal carbon footprint.

Money is arguably a good measure of your worth to the community as a whole: for a statistician, a Judgement Day starting point would be the average monetary value of a life. Your life really does have a finite monetary value: if life were priceless, you wouldn't travel in a car, plane, boat or train unless it is 100% safe.

A zero-risk plane or train would cost an infinite amount of money – which is another way of saying that it's impossible to make anything completely safe. Typically, doubling the amount of money spent on safety systems will decrease the risk of failure by a small percentage. Doubling the safety spend again will lead only to a similarly small reduction in risk.

On the other hand, we wouldn't tolerate public transport on Heath Robinson trains and jalopies that had been built for next to nothing. The reason: our lives are worth more than the cost of improving the safety of the train.

So we reach a compromise: we spend a lot of money to make transport systems, industrial plants, and so on reasonably safe, but the amount spent is related to the perceived monetary value of a life. We don't spend too much, and we don't spend too little.

There's another uncomfortable reality: the monetary value of a human life depends on the wealth of the community in which that life exists. In Western countries we spend a lot to make life fairly risk-free. It's a reflection of our wealth and ultimately of the value we place on our lives. But move to a poorer country – Bulgaria, say – and it's unlikely you'll find the cars, trains and planes as safe as in Britain. It's simply because the monetary value of a life in Bulgaria is not as great as it is in Britain.

The value of your life also depends on who is picking up the bill. We demand high safety standards in a nuclear power station: if you lost your life because of a nuclear accident, you would expect your family to receive a hefty financial payout from the operator. If, though, you climb into your car and floor the accelerator, the risk of you dying is hundreds of thousands of times greater than death by nuclear accident. But you have made the choice to risk your own life; it's your life and you can dispose of its value as you please.

So, how much is your life worth? For a rough guide, your life insurance policy will tell you the most that the company will pay out if you die: my policy promises £1.5 million. Or, look in the newspapers to see how much compensation is paid to the families of victims of disasters such as the Paddington and Hatfield train crashes. All these amounts are in the same ballpark. In fact, companies running public transport, power stations or industrial plants do quantitative risk assessments on their business and put the value of a human life into the cost–benefit equation. It's reputedly around £3 million.

In the entries in the *Book of Your Life*, the monetary value of your life might be equal to the net profit we made from birth till death. The average annual income for a male working British adult is about £24,000. Work for forty years and you'll have made £24,000 × 40, about £1 million. Yes, most of us will nearly be millionaires by the time we stop working. The trouble is that the money trickles in over your working life rather than arriving as a lump sum. Notice that it's comparable to the payouts made by insurance companies for a death, again indicating that your life has a monetary value that reflects your total earnings.

But, as for any business, your life accounts should show expenses as well as income. As far as the community is concerned, your worth is the net amount by which your birth and life increase the community's wealth. On this basis, the value of your life is the amount you earn minus the amount spent on keeping yourself alive so that you can earn. Those expenses include food, shelter, education, health care and pension contributions.

We generally pay for food and shelter ourselves. Take shelter: the average price of a home costs around £150,000 and is shared by two people, so shelter for one person for life costs around £75,000. That's assuming the house lasts for about one lifetime.

Then there is fuel to keep you warm and for cooking. Add your electricity, gas and oil bills for a year and multiply by the eighty years of expected life. It'll come out at around £500 per person per year, making your lifetime fuel bill £40,000. And of course there is water and sewage disposal: that's another

£250 per person each year. Twenty thousand pounds for a lifetime should cover it.

Food is amazingly cheap in the developed world. The weekly bill for essential food for one person in Britain is about £20 – though many of us spend considerably more on all sorts of delicacies that are not really needed. If you live for eighty years, you'll have had 4,000 weeks of essential food: a total bill of £80,000. You might double that figure to £160,000 if you have gourmet tendencies.

The expenses so far total £295,000 – most of us are gluttons if not gourmets. Realistically, you'll also be spending money on frivolous luxuries like clothes. That's going to push your expenses up by at least £150,000.

Transport is very expensive. It's largely used for going to work and getting the shopping, so it's not a frivolous form of consumption, and you might do it by bus, train, plane or car. If you choose to use a car, perhaps jointly owned with your partner and driving 20,000 miles per year, you'll have an annual expenditure of £2,000 on fuel, around £500 on depreciation and maintenance, and £500 on tax and insurance: a massive £3,000, or £1,500 per person. That's £120,000 in a lifetime – one and a half times your essential food bill.

And you still haven't paid for education, pension, health and your security. These items are paid mainly through taxation, and at current rates someone earning £24,000 will pay nearly £5,000 in income tax and council tax. Deducted over forty years of your working life, that'll amount to another £200,000 of expense.

So, the financial section of the *Book of Your Life* is going to look something like this:

Item	Income	Expenses
Earned income	£1,000,000	
Food		£160,000
Shelter		£75,000
Heating and cooking fuel		£40,000
Water		£20,000
Clothes, etc.		£165,000
Transport		£120,000
Health, education, etc.		£200,000
Sub-totals	£1,000,000	£780,000
Net profit	**£220,000**	

This bit of your ledger isn't looking too bad. But, of course, the wealth you generate over forty years physically decays. It's the Second Law of Thermodynamics applied to your nest egg. The entropy of any closed system increases over time; the fallen log rots away, the bricks in your house weather and turn to dust, the car engine wears out.

All the *real* wealth you will have generated in a lifetime is physical: you can't eat £5 notes. Real wealth is bricks and mortar, machines and surplus food. Its value has an abstract monetary representation, but wealth is real physical stuff. And it all decays with time.

The value of hi-tech goods vanishes rapidly – computers, for example, are usually discounted to zero over just four years. The value of a car reduces by about 50% every two years. Bricks and mortar are a better investment: the effective life of a domestic house is around fifty years. That's not to say that your typical house rots away in fifty years, just that, on average and without maintenance, the components of a house last fifty years.

At a conservative guess, most of the wealth you created at the beginning of your working life will have been dissipated by the end of your life, while the wealth created at the end of your life will be undiminished. As a result, the difference in the communal wealth at the end of your life and the wealth that would have been if you had never been born is probably no more than half of the £220,000 suggested by the raw numbers.

So, it looks as if the real economic value of your life from cradle to grave is only around £110,000 – not much more than £1,000 for each year of your life. But at least the balance is positive: you won't be cast into the Lake of Fire for this bit of the *Book of Your Life*.

But look carefully at the words from the Book of Revelation with which we started. It says 'according to their *works*' – that's plural. What one does for money is not the only kind of work. Surely selfless piety ought to count for quite a lot in the credit columns. Whose life has the greater worth, Mother Teresa's or Rupert Murdoch's?

Rupert Murdoch has created a business empire which employs tens of thousands of people. Each of those employees is generating wealth because of Murdoch's skill in creating and maintaining the organisation. Without News Corp., the average income in Britain might be slightly less than its current level. If it were just £10 less, we could argue that Mr Murdoch is responsible for creating an annual wealth of £10 multiplied by the working population, say £200 million. That's a lot more than any ordinary individual is going to create in a lifetime.

Mother Teresa founded and maintained a religious order

that has cared for thousands of orphans, disabled and sick people. How should we value acts of kindness, which we might crave ourselves but find hard to give: making friendly eye contact with a down-and-out on the street, or giving a word of encouragement to an insecure child?

It might not seem possible to place any monetary value on that kind of 'work'. But kindness, compassion, respect and care are behavioural traits – traits that hold people together, allow social organisation, and ultimately facilitate the economic activity that drives our modern industrial world. They have economic value; without them, News Corp. wouldn't function. They are just difficult to work out.

SPEAK GEEK

A DONATION OF 1P EACH YEAR FROM EVERY
WORKING PERSON IN BRITAIN WOULD PROVIDE YOU
WITH AN AVERAGE SALARY.

Out of a population of over 60 million, there are only about 25 million people in Britain who are actually working on any given day. If each of them were to send 1p to your bank account you would get 25 million pence, or £25,000. Twenty-five million annual direct debits for 1p would be sufficient to provide you with the average national income.

Make those direct debits monthly, and you'd be a fat cat. Now that's the kind of benefit I like.

27
HEAVENLY BODY

How much does the Moon weigh?

The Moon weighs about 75 billion billion tonnes. I know this because the width of my little finger at the third joint is 10 mm.

I once lived on a university campus in a row of terraced houses. We were bracketed by delightful neighbours who became lifelong friends. On one side was a neuroscientist who had voluntarily inverted his working day. Every evening, at about 9 p.m., I heard him start his car as he left for the 500-metre journey to his research laboratory, and at 3 a.m. every morning I would be woken as he arrived back.

On the other side lived a mathematician. His route between home and office was determined by the principle that short cuts are a heresy against an ordered world. Each evening on his return from work, he walked home along the opposite side of the street. Arriving at a point precisely opposite his house, he would turn sharply through 90 degrees, and cross the road to his front door. Not for him the licentious joy of a walk along the hypotenuse.

Sandwiched between these archetypal academics, we lived very ordinary lives – apart from the grinding noise made by my telescope-making machine at the top of the stairs.

Some telescopes (and all large ones) use a concave mirror rather than a glass lens to gather light. They are called reflecting telescopes. Like a prism, a lens splits light into different colours. An image viewed through a cheap lens telescope – a refracting telescope – will show blurred colours on the edges of objects in the field of view. A mirror doesn't split light in this way, so the image in a reflecting telescope is free from unwanted coloured fringes.

The simplest kind of reflecting telescope is called a Newtonian telescope. It has a concave mirror at the bottom end of the telescope tube. Light from whatever is being observed goes down the tube and is reflected back up to form an image somewhere just near the top end of the tube. A second mirror, small and flat and mounted at 45 degrees, interrupts the light path and directs the light path out of the side of the tube where the image is viewed through a magnifying lens – the eyepiece. A reflecting telescope is actually just a microscope – to look at the image – coupled with a large mirror to gather and concentrate the faint light from the night sky.

The surface of the curved mirror has to be highly polished. Most important, and this takes a lot of care and effort to achieve, the profile of the curve must be highly accurate. But, surprisingly, you can make a good telescope mirror in your own home, using your own hands and some basic equipment.

It's done by rubbing two circular discs of glass together in a sandwich filled with abrasive paste. The discs will wear into a convex mirror and a perfectly matching concave mirror. The mirror profile will be accurate to a few millionths of an inch, thanks to the unavoidable consequence of geometry, wear and force between the two pieces of glass.

There is real excitement in looking at the Moon and planets through a telescope, especially if you've made it yourself. Being outside on a frosty night and knowing that the image projected onto your retina is made by the impact of photons of sunlight that bounced off Saturn or Jupiter an hour or so earlier gives a sense of perspective on life.

The stars are all so distant that they remain points of light regardless of how much your telescope magnifies their image. By comparison, the Moon is so near to the Earth that a telescope isn't needed to see detail on its surface. Here's how to estimate the weight of the Moon, using just your finger and no telescope.

Go outside on the night of a full Moon and raise your index finger in front of one eye until it covers the Moon. Move your finger nearer or farther from your eye until the width of your finger just covers the Moon's diameter. At that point, make an estimate of the distance of your finger from your eye. Now go back inside.

The Moon is about 400,000 km from the Earth. I found that my 10 mm wide little finger had to be held at arm's length to cover the Moon. That's roughly 1 metre.

Now imagine two pieces of string (you'd need quite a few balls of it) attached to either side of the Moon and stretched back taut to your eye. The strings meet at your eye, but one metre back towards the Moon, the pieces of string are already 10 mm apart. Go another metre and they would be 20 mm

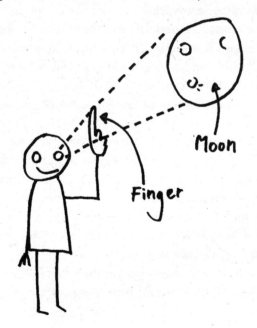

Moon

Finger

apart. Every metre along the string takes them another 10 mm apart. Go 400,000,000 metres, and you will be at the Moon. The strings will then be separated by 400,000,000 times 10 mm. That's 400 million multiplied by one-hundredth of a metre. So the Moon has a diameter of 4 million metres. This estimation technique, based on *similar triangles*, can be applied to virtually any distant object as long as you know how far it is from your eye.

Actually, our estimate of an arm's length was over-generous at 1 metre; it's more like 85 cm. Our estimate of 4 million metres should therefore be reduced by 15%, to 3.4 million metres, or 3,400 km.

Back to the original question, which was to find the weight of the Moon. We've estimated its diameter, and now it's just a short step to get its weight. First work out the volume, then estimate the weight of 1 cubic metre of moonrock, and finally multiply the two numbers together to get the weight.

The volume of the Moon – and of any sphere – is one-sixth times π times its diameter, times its diameter, times its diameter. This formula is written as:

$$\text{Volume} = \tfrac{1}{6} \times \pi \times \text{Diameter}^3$$

The mathematical constant π is about 3, so we can take the $\tfrac{1}{6} \times \pi$ in the formula to be $\tfrac{1}{2}$, and the volume of a sphere to be roughly half the cube of its diameter. (Think about it: the volume of a sphere is about half the volume of a box into which it would just fit.)

Applying this rough and ready formula to the Moon gives its volume as:

$$\tfrac{1}{2} \times 3,400 \times 3,400 \times 3,400 = 19,700 \text{ billion cubic kilometres}$$

That's near enough 20,000 billion cubic kilometres. Now estimate the average weight of a cubic metre of moonrock. We'll assume that moonrock has a similar density to Earth rocks. I've visited a granite quarry where the rock is cut into slabs about 2 metres long, 60 cm wide and 60 cm thick. Each slab weighs about 2 tonnes and their volume is about two-thirds of a cubic metre. That gives 3 tonnes per cubic metre, or 3 billion tonnes per cubic kilometre.

Putting this rock density together with the Moon's estimated volume of 20 billion cubic kilometres gives a weight of 60 billion billion tonnes. The presently accepted value is about 74 billion billion tonnes, so the estimate turns out to be quite good considering it's based on judging the width of your little finger.

SPEAK GEEK

SUNLIGHT IS 400,000 TIMES BRIGHTER THAN
MOONLIGHT.

Light from the Sun expands outwards into space. After 93
million miles, a tiny fraction lands on the Moon, from which
some of it is scattered back into space as moonlight. The total
amount of scattered moonlight is determined by three factors:
the area of the Moon facing the Sun, the Moon's distance from
the Sun, and the reflectivity of the Moon's surface. A small
fraction of that moonlight lands on the Earth's surface after a
journey of around 250,000 miles from the Moon.

The intensity of the Sun's light that arrives on Earth via
reflection off the Moon – moonlight – works out at about
$2\frac{1}{2}$ millionths of the brightness of light coming directly from
the Sun itself. In other words, sunlight is 400,000 times as
bright as its more romantic cousin, moonlight.

28
PASSING WATER

How many times has the River Thames been through the human body?

I was once taken by a student to his workplace: the sewage processing works on the edge of a provincial town. It was a sparkling, breezy autumn morning as we walked across one of the steel gantries spanning the filter beds. The morning was so sparkling and breezy that spray from nozzles on the rotating arms over the beds made a rainbow in the air above us. Drops of distinctly second-hand water were being blown onto my face.

Unaware of my single-minded determination not to open my mouth or lick my lips, the student enthusiastically ushered me to the final outflow from the farm where the purified effluent was discharged into the local river. We leant over the parapet and watched as electronic sensors continuously monitored the small but fast-flowing river generated by a town with a population of some 100,000 people. The sensors signalled to a computer that the effluent was now pure enough to drink. But I wasn't thirsty.

There's an urban myth that water in the River Thames passes through a human body seven times before it leaves London for the sea. The basis for this assertion is that water is extracted from the Thames for supply to Londoners and people living in the Thames Valley, who drink it, cook with it and make tea with it. The water then returns via sewers to the River Thames where it is extracted to start its next circular journey via the human bladder.

There may be something in the story. In 1852 an Act of Parliament was passed making it an offence to extract water from the tidal Thames in London. The Act aimed to protect Londoners from disease, chemicals and sewage carried by the river. As a result, two big water extraction plants were built.

They still operate, upstream in the non-tidal part of the Thames, isolated from London sewage by the last locks on the Thames, at Teddington and Richmond.

Getting enough clean water to people, even in rich and technologically developed countries in Europe, is difficult. It will get harder as populations increase, and harder still if the climate change projected by some scientists comes about, giving us very hot summers and uneven rainfall patterns. Water shortages might then force populations to migrate northward. At best it will cause economic damage; at worst it will lead to war.

One proposal is to augment London's water supply from a desalination plant sited downstream of London in the Thames estuary. There is no shortage of dirty salty water in the Thames below London: it's effectively part of the North Sea. But any desalination plant requires heat energy from gas, oil or electricity to boil the dirty water and then condense the steam to form pure water.

Boiling water takes a lot of energy. First, the water temperature has to be raised from the river temperature of around 10 degrees Celsius to 100 degrees. Heating 1 gram of water by 1 degree takes 4.2 joules of energy. A 10-litre bucket of river water weighs 10,000 grams, so it'll need 10,000 × 90 × 4.2 = 3.78 million joules of heat energy. For comparison, a good gas camping stove will give about 2,000 joules of heat energy each second, so it would take 3.78 million/2,000 seconds to heat the bucket of water. That's just over half an hour.

But the water won't boil just by raising its temperature to 100 degrees. Boiling with bubbles of steam only starts when the temperature rises a tiny bit above 100 degrees, and making this happen requires an enormous amount of extra heat; much more than was used to heat the cold water to 100 degrees.

The extra heat to make the water change to steam bubbles is called the latent heat of evaporation and is equal to 2,268 joules for each gram of water changed to steam. Changing a 10-litre bucket of 100-degree hot water entirely into steam will take a further 10,000 multiplied by 2,268 = 22.68 million joules.

That will take your camping stove another 3 hours and if you used an efficient large-scale burner like the one in an oil-fired central heating system, it would use about three-quarters of a litre of heating oil. Three-quarters of a litre of oil for a bucket of pure water doesn't sound too bad – until you multiply it by the number of buckets of water used each day.

If you are an average Brit, you'll be using about 160 litres of water each day. That's sixteen big buckets. Apply that to a provincial town of 100,000 people, and the domestic water consumption is 100,000 × 160 = 16 million bucketfuls per day.

To boil each bucket of water would require three-quarters of a litre of oil or fuel equivalent, giving a daily oil burn of 11.2 million litres – that's like having your central heating oil tank refilled 7,000 times each day. Not surprisingly, the Thames desalination proposal is mired in controversy, not least because of its very high energy requirements.

But how much of that dirty Thames water has really passed through human bodies? To get the answer, we need to know how much water passes through a single human in a day. We'll also need to be a bit more precise about who is consuming the water and which bit of the River Thames we're talking about. A convenient boundary is Teddington, where the River Thames becomes tidal. We'll do our calculation for the water and people upstream of this point. (A calculation on the river downstream of London wouldn't be very meaningful because the water flows in and out with the tide twice a day.)

The amount of effluent produced by each person in a day will be similar to the amount of fluid they take in by mouth. Of course, some will be lost as perspiration, but a large part will be disposed of in the toilet pan. You could look up the daily amount of water drunk or taken with food by a typical person, but it's easy to examine your own habits and arrive at the answer. I recorded my personal oral fluid consumption for a day and got this list:

Time	Description	Intake method	Volume
Breakfast	Tea/coffee	2 cups	0.50
	Juice	1 glass	0.25
	Milk	With cereal	0.25
Mid-morning	Tea/coffee	2 cups	0.50
Lunch	Water/juice	1 cup	0.25
	Water	In cooked food	0.25
Afternoon	Tea/coffee	2 cups	0.50
Evening meal	Water	In cooked food	0.25
	Juice/milk/water	1 glass	0.25
Evening	Tea/coffee	2 cups	0.5
	Social beer/wine	2 glasses	0.5
Total water consumption			**4.0 litres**

Just four litres passed through my body. You could live quite happily on four litres per day, but you actually have about 160 litres delivered to your house through the water company's pipes. If it were left on your doorstep with the morning milk, there would be about 300 pint bottles next to the milk, every day. The 156 litres of water that doesn't pass through you is used for flushing the lavatory, washing clothes and dishes, and a host of other essential and not-so-essential water purposes.

To find the total amount of water added to the river after passing through a human body, we need to know the number of people living in the Thames Valley, say upstream of Teddington Lock. It's safe to assume that the effluent from anyone living in the Thames Valley area ends up in the Thames – effluent will not flow uphill, out of the valley, of its own accord. A simple approach is to look at a relief map of Southern England and draw a line along the high ground to the north and south of the Thames, and identify the big centres of population lying between the lines.

The water companies serving the region have about 8 million customers, but quite a lot of those will be in London itself, below the tidal lock at Teddington. But there are large towns and cities upstream of the lock: Oxford, Reading, Slough and High Wycombe, for instance, all drain their waste water into the Thames.

We could take the population in the Thames Valley above Teddington to be roughly a million people. To get an exact value you'd have to use the relief map and make a detailed count of all the major towns lying in the valley. It could be 0.5 million and it could be 4 million, but we are just interested in getting an approximate answer, so a round figure of 1 million will do. The amount of water flowing daily into the Thames that has passed through a human body is therefore 1 million multiplied by 4 litres per person: 4 million litres.

It's more convenient to change that figure into litres per second. There are 3,600 seconds in one hour and 24 hours in a day, so in one day there are $3,600 \times 24 = 86,400$ seconds. To get the human water flow in litres per second, just divide the daily rate by this number. That's $4,000,000/86,400 = 46$ litres per second. It sounds a lot, but how does it compare with the total flow in the non-tidal part of the Thames?

If you have ever been out in a punt or a rowing boat on the Thames at Kingston upon Thames or nearby, you'll have found that there is a gentle current in the centre of the river, of about 3 kph. That's about a metre per second. It's easy to confirm this by dropping a stick in the river and timing how long it takes to travel, say, 10 metres.

The river at Kingston is about 2 metres deep in the middle and about 50 metres across. If the depth were the same all the way across, the cross-sectional area of the river would be $2 \times 50 = 100$ square metres. In reality, the river becomes shallower at the edges, making the cross-sectional area less – let's say 50 square metres.

The current is also much less near the bottom and the banks of the river than at the surface in the middle, so let's take the average current across the river to be half our initial estimate. That's 0.5 metres per second. On a rainy day the current will be much stronger because of all the water running off the surrounding land, but we'll assume that the river is in a hot, dry summer state.

If all the water in the river is flowing downstream at 0.5 metres per second, and the cross-section of the river is 50 square metres, then in one second the volume of water that passes a given point on the bank is $50 \times 0.5 = 25$ cubic metres. So, the flow rate of the non-tidal River Thames is about 25 cubic metres per second. One cubic metre of water is 1,000 litres, so another way of stating the flow rate is 25,000 litres per second.

Our original question can now be answered. The average flow rate of water through the bodies of all humans living upstream of Teddington is about 50 litres per second, but the flow rate in the river is 25,000 litres per second. So only one-five-hundredth of the water flowing past Kingston upon Thames has actually seen the inside of a human bladder within the last few days. Of course, over a period of millennia, it's likely that every drop of water in the river has seen the inside of someone's or something's bladder, but you probably won't be too worried by a biological contact that happened in the Jurassic Period.

But hang on – there's the other 156 litres of water used by each person every day. A significant amount of that is used to flush the lavatory. It's about 10 litres for a full flush, so you probably flush 50 litres of water down the loo each day. Then every time you have a bath you'll use another 50 litres. Your water consumption is already 100 litres a day, even without washing your clothes and dishes, not to mention the extravagances of car washing or watering the garden.

Much of that water from those activities is contaminated by sewage or detergents, and after purification at the kind of sewage farm I visited with my student, it finds its way through small rivers back to the Thames.

If we count the total volume of used water that's returned to the Thames, the myth looks a little bit closer to reality. One million people, each consuming 160 litres daily, gives a total of 160 million litres. That works out at around 2,000 litres per second, which is almost a tenth of the water flowing past Kingston upon Thames.

Remember this when you go swimming. Ten per cent of the water around you has been through a lavatory, dishwasher or washing machine. It is a remarkable feat of engineering and organisation that for the past twenty years this water has been so well purified that it presents a negligible risk to your health.

SPEAK GEEK

A THUNDERCLOUD CAN HOLD MORE THAN 1,000 TONNES OF WATER.

All the water we will ever consume has spent time as rain-clouds. But how much water can a cloud hold? A true Geek will have a go at working it out, but will also recognise that the estimate could easily be ten times too big or ten times too small.

Cloud thickness ranges from hundreds to thousands of metres; for simplicity, let's go for a 1,000-metre thick, 1,000-metre long and 1,000-metre wide cumulus cloud. Multiply them all together and you've got a total volume of 1 billion cubic metres, or 1 cubic kilometre.

So how much does a cubic metre of cloud water droplets weigh? The other day, I heated some water inside a freezing cold shed. After 50 ml had boiled away, inside the shed was much like being on a mountain-top in thick mist. The volume of my garden shed is 50 cubic metres, so the density of water droplets was around 1 ml – that's 0.001 litres – per cubic metre. So, maybe the cloud contains 1 billion × 0.001 = 1 million litres of water. That's 1,000 tonnes to you and me.

29
THE MAN IN
WHITE

How much could sea levels rise?

In 1959 I was lying in a bed in Great Ormond Street Hospital for Children. It was early evening. Matron appeared, accompanied by a nurse. She checked that the area around my bed was clean, briskly straightened the bedclothes and then, with a stern face, brushed my hair. Curtains were pulled quickly around the bed and then she was gone. The nurse stayed to whisper, comfortingly, that the *Consultant* was coming.

The curtains were pulled aside, and there he was, surrounded by a group of respectful trainee doctors, each in a white coat and wearing their stethoscope round their neck. The Consultant was *Mr*, and he was the only person not wearing a white coat.

Even to an eight-year-old, he was beyond question The Man Who Knew Everything. His words were The Truth.

That encapsulates how we handle complex bodies of knowledge and skill. We hand over the responsibility for knowing and doing to trusted individuals or groups, just as priests were once believed to hold special knowledge and were therefore entrusted with making our connections with gods and the elements.

Often, those trusted individuals have special regalia to mark them out – the equivalent of a priest's robes. Doctors and scientists wear white coats, the academic professor wears a bow tie, the judge wears a wig. Even a fluorescent orange safety jacket is a signifier of authority.

But handing over responsibility for identifying truth to doctors, scientists, journalists or whoever is dangerous unless you are able to make basic logical checks on their pronouncements. If you don't already doubt professional wisdom, just think about the number of changes there have been over the years in professional advice on which way up to place your baby in its cot or the safety of lead additives in petrol, or the scientifically flawed study on links between autism and the triple vaccine MMR.

In medicine, the kind of sanity checks we can all carry out without possessing expert knowledge are usually simple statistical questions. How many people took part in the trial of a new therapy? How were they selected? Is the reported effect causal or just associative?

In the environmental sciences, the tests are to do with rudimentary, common-sense physics. Anyone can check the logic of expert pronouncements. For example, take this imaginary newspaper headline:

ARCTIC SEA ICE MELTS: MASSIVE FLOODING
THREATENED AS SEA LEVEL RISES

It's obviously true: everyone knows that ice is made of water, and if melted, that water is added to the ocean and will raise the sea level. But what kind of ice are we talking about?

Most of the Arctic ice visible from space is sea ice. It is floating in the water of the Arctic Ocean, and if floating sea ice melts it will make absolutely no difference to the volume of the ocean, or to its level. The Greek philosopher Archimedes pinned this down in what we now call Archimedes' principle.

The gist of his famous Eureka! moment is this. Anything that floats – a ship, a twig, a swimmer – displaces a volume of water that is exactly equal to its own weight. A floating iceberg is already displacing water with its weight. Melting the iceberg doesn't change the weight of water that it originally contained, so the sea level won't change.

If you want to test Archimedes, do an experiment in your kitchen. Put half a dozen ice cubes into a plastic measuring jug. Top up the jug with water to, say, the 500 ml mark, and wait. The ice cubes will float in the water; with about 10% of them above the water and 90% below, just like a real iceberg. When they have completely melted, check the level of water in the jug. If it has risen, you will have confounded a two-thousand-year-old scientific principle. If it hasn't, you will have proved that melting sea ice will not cause flooding.

Here's a similar sounding but physically very different headline:

ANTARCTIC ICE MELTS: MASSIVE FLOODING
THREATENED AS SEA LEVEL RISES

Now, unlike polar ice at the North Pole, the ice at the South Pole is resting on land: on the continent of Antarctica. If this ice melts, it really will add water to the oceans. But how much will the sea level rise as a consequence?

We could take a worst-case scenario and work out how much the sea level would rise if all the Antarctic ice melted, however long that took. To work this out we need two pieces of information: the volume of water in the ice resting on the continent, and the area of the world's oceans. Dividing the volume of melted ice by the ocean area will give the approximate rise in sea level. (The actual rise will be slightly less than calculated because land will be flooded as the sea rises, and a very small amount of the water from the melted ice will be used to fill this new bit of ocean volume.)

To picture how the method works, imagine a garden pond whose area is 10 square metres. Imagine also that you have a cube of ice of volume 1 cubic metre. First melt the block of ice. It will shrink a bit as it melts because ice takes up more room than water, and you will end up with about 0.9 cubic metres of water. Now pour the water into the pond. The volume of the added water will raise the water level by $0.9/10 = 0.09$ metres, or 9 cm.

The method seems to work, so let's apply it to the oceans. First, what is the volume of ice on the Antarctic continent? That's worked out by multiplying the area of the continent by the average ice thickness. You could look up these figures; the area is about 14 million square kilometres, and the average ice thickness is 1.6 km. That gives a total ice volume of 14 million $\times$ 1.6 = 22.4 million cubic km. The ice will shrink as it melts, and the water volume will be about 90% of the ice volume; 90% of 22.4 million cubic km is, near enough, 20 million cubic km.

Now we need the surface area of all the Earth's oceans, to which the melted ice water will be added. About 70% of the Earth is covered by water, and the total surface area of the oceans is around 360 million square km. Dividing the melted ice volume of 20 million cubic km by 360 million square km tells us how much the sea level would rise if all the Antarctic ice melted. It's 20/360 = 1/18 km.

That's 55 metres. So, if all the Antarctic ice melted, much of London would be submerged. You could probably sunbathe on a beach near the top of Parliament Hill or Hampstead Heath while watching the waves lap around the lower quarter of the NatWest Tower, or the buildings at Canary Wharf.

This is a pessimistic calculation. It's very unlikely that all that ice is going to melt any time soon. There is even evidence that ice cover in parts of Antarctica is on the increase. But we do seem to be in a period of global warming which could cause sea levels to rise, and this can happen without any ice melting anywhere. Imagine another news headline:

OCEANS HEAT UP: MASSIVE FLOODING THREATENED
AS SEA LEVEL RISES

Water expands as it warms up. The volume of the water in the oceans will increase as they are warmed, and that will make the sea level rise. This will happen regardless of any increase in volume caused by ice melting.

Although it's one of the most familiar substances of all, water is quite strange in several respects. One of its peculiarities is the way in which its density changes with temperature. At just above freezing point, one kilogram of water has a volume of 1.0002 litres. Increase the temperature to 4 degrees Celsius, and the same kilogram shrinks to almost exactly 1 litre. Above 4 degrees its volume starts to increase again, by roughly 0.04% for each degree of temperature rise.

The figure of 0.04% volume rise per degree gives us a way

of working out how much the sea level would rise as the water expanded. We need to know the total volume of the Earth's oceans, the probable rise in temperature of the water due to global warming, and the oceans' total area.

The total area of the Earth's oceans is 360 million square km. Their combined volume is the average depth multiplied by the area. Living on an island on the continental shelf, we are used to sea depths of less than 200 metres, but the average depth of the Atlantic is 3.6 km. The Pacific is even deeper: 4.3 km, on average. The Indian and Southern Oceans are similar. In fact, the deepest point of the Pacific Ocean is 11 km, about the same distance below the surface as jet airliners fly above it. If you sat at the bottom of this ocean trench and looked up, a jumbo jet floating on the surface (if you could see it, which you actually couldn't) would look the same size as those pinpricks of silver you see in the sky.

A fair estimate of the average ocean depth over the whole globe is 4 km. With an area of 360 million square km, the volume works out at 4×360 million = 1,440 million cubic km. Raise the temperature of the whole ocean by 1 degree, and its volume will increase by 0.04%. That's an increase of 1,440 million $\times$ 0.0004 = 576,000 cubic km.

The increase in volume will be spread over the entire area of the Earth's oceans. The resulting sea-level rise caused by a 1 degree increase in global temperature will be 576,000/360 million km: about 1/600 km, or 1.7 metres. According to the gloomiest predictions, the global temperature will rise by about 4 degrees over the next century, in which case we could expect the seas to rise by $4 \times 1.7 = 6.8$ metres, simply as a result of their expansion.

But how soon is this going to happen? Should you think

about moving to a higher area before house prices rocket? Here's another headline:

EARTH IN MELTDOWN: ANTARCTIC ICE WILL
DISAPPEAR IN FIFTY YEARS

The rate and causes of global warming are still fiercely disputed. Most climate models point to increased atmospheric carbon dioxide, possibly coupled with a natural periodic increase in the Sun's power as a cause of global warming. The mechanics of our biosphere – the atmosphere, ocean, surface environment and life viewed as a single system – are so complex that no simple back-of-the-envelope calculation will tell us whether the current predictions of climate scientists are correct.

Still, you can attempt to test that last headline about all the Antarctic ice melting within fifty years. An important and obvious cause of ice melting is direct heat from the Sun. As the ice melts, particles of dirt that were trapped inside the ice start to build up on the surface, which thus grows darker. Dark rocks are exposed at the edges of the melting ice sheet. The darkening surface absorbs more of the Sun's heat and reflects less back into space. This further hastens the process of melting, leading to an unstable positive feedback effect.

Some climate scientists believe that this is what they call a tipping-point mechanism of climate change that might lead to irreversible melting of the ice sheet. There are processes other than direct solar heating that might cause the ice to melt, for example hot winds from temperate regions, or simple proximity to an Antarctic atmosphere whose temperature has been raised above freezing by global warming. New snowfalls would tend to cover up the exposed dark surface of the melting ice to make it more reflective and counter the melting, but overall the melting processes might be overwhelming.

Can we work out how much heat energy would be needed to melt the entire Antarctic ice sheet? We have already found that its volume is about 20 million cubic km, and that's a weight of about 20 million billion tonnes. If the amount of heat needed to melt 1 tonne could be calculated, we'd be able to work out the energy needed to melt the lot.

One gram of ice needs 336 joules of heat to change it from ice to water. That's the heat you would get by running a 3 kW electric heater for about one-tenth of a second. To melt 1 kilogram of ice you'd have to have the heater on for 1,000 times as long – for about 100 seconds.

You can do a simple check on this figure next time it snows: empty your electric kettle and then stuff it full of snow. Switch on and time how long it takes until all the snow has just melted. Then weigh the water. For every kilogram weight of water, the kettle should have taken about 100 seconds to do the melting.

If – and it's a big if – the energy necessary to melt the ice in Antarctica arrives mainly as heat from the Sun, we need to know how much heat energy the Sun delivers to the Earth's surface. We can start at London on Midsummer's Day. At noon, if the air is very clear, the Sun will throw around 1,500 watts of heat onto each square metre of the ground. That's about half the heating effect you would get from standing in front of an old-fashioned radiant electric heater. On a cloudy day in January it's only about 20 watts per square metre.

At night, of course, there is no incoming heat. In fact, the process is reversed: if the air is clear, heat is radiated back into space during the night. That's why there is sometimes a ground frost, even when the air temperature is above zero. Overall, it is estimated that the average power of the solar heating in London is 100 watts per square metre. That's averaged over twenty-four hours a day, all year round.

At midsummer in London the Sun reaches an altitude of about 61 degrees, but in Antarctica it is much lower in the sky, and its power much less. In fact, from the South Pole the Sun's altitude is only 22.5 degrees at noon on Midsummer's Day. That's only a little more than in London at noon in midwinter.

Based solely on the angle of the Sun and the lengths of the days throughout the year relative to London, the average solar heating at the South Pole will be about 40% of the London value: about 40 W on every square metre. There are other factors to take into account, such as the greater thickness of the atmosphere through which the low-altitude Sun's rays have to travel in Antarctica. But the atmosphere may be much clearer in Antarctica than over London, so we'll keep 40 W per square metre as our final figure.

The crux of the prediction of catastrophic melting made by some climatologists is that the melting ice will leave dark dirt and debris on the surface of the ice, lending it a dark colour which will absorb more of the Sun's rays instead of reflecting them.

Clean snow can reflect about 90% of the incident rays, and even dirty snow and ice will reflect some of the heat. The exact amount will depend on how dark the ice and snow become. We'll assume a figure of 75%, although actual measurements are needed to find the true value. That would mean that 25%

of the Sun's heat is absorbed by the dirty ice and causes melting, which gives us about 10 W per square metre.

We have already calculated that it takes a 3,000 W heater about 100 seconds to melt a kilogram of ice. The Antarctic sunlight's 10 W per square metre of absorbed power will take $3,000/10 \times 100 = 3,000$ seconds to melt 1 kilogram. Very roughly, that's 1 kilogram melted every hour in each square metre of Antarctica.

A kilogram of ice has a volume of a little over 1 litre, and when it's melted it will form a layer of water 1 millimetre thick over the 1 square metre area. So, it looks as if the Sun might melt 1 mm of ice every hour. With around 10,000 hours in a year, the yearly melt depth would be 10 metres. It sounds a lot, but the average ice sheet thickness in Antarctica is 1.6 km, and, with our assumptions, it would take 160 years to melt completely.

So, our imaginary headline

EARTH IN MELTDOWN: ANTARCTIC ICE WILL
DISAPPEAR IN FIFTY YEARS

is almost certainly wrong.

In fact, this last simplistic estimate shows that sometimes experts are just what we need: more sophisticated predictions by climate scientists and geophysicists put the time to melt the ice sheet at nearer a thousand years. That's still potentially very serious, but not imminent.

Thank goodness for experts – sometimes.

SPEAK GEEK

NINE NEW TREES NEED TO BE PLANTED EACH YEAR TO OFFSET YOUR CO_2 EMISSIONS.

Wood is composed mainly of a substance called lignin cellulose. A fully grown pine tree contains up to half a tonne (500 kg) of the stuff. About 44% of the weight of each molecule of cellulose is made of carbon, so the tree contains 220 kg of carbon. That 220 kg will have come from just under 1 tonne of atmospheric CO_2 while the tree was growing.

The average Briton is responsible for 9 tonnes of CO_2 emission each year. In that case, you'll need to grow nine new trees each year to be carbon neutral. In Britain, this would mean planting and maintaining 540 million new trees every year. The spacing between mature pines in a forest is 4 to 5 metres, so each tree occupies a ground area of around 20 square metres. In round figures, the area of land needed for new forest will be 10 billion square metres. That's one new forest 100 km long and 100 km wide planted every year.

Britain would be completely covered in forest in less than 15 years. You wouldn't see the wood for the trees.

ACKNOWLEDGEMENTS

This book wouldn't have been started or finished without the encouragement and support of my wife, Seija.

The seed of the idea wouldn't have been presented to the publishers if my agent, Cathryn Summerhayes, hadn't recognised that this was a book worth writing and helped enormously in the development of the initial drafts.

And, once the writing started in earnest, my editors, Ben Dunn and Jack Fogg, provided constant and invaluable guidance about what works and what doesn't, as well as helping make dry calculations into something more fun.

My friends Mike Carey, Ramsay Gibb, David Vincent, and Richard Aldridge all helped me with their own personal contributions, and John Woodruff provided the final painstaking and tolerant editing.